Contemporary Rural Geographies

This volume draws together a set of original research essays covering a range of themes in the rural geography of contemporary Britain by leading British rural and environmental geographers. The chapters also mirror Richard Munton's research path and have been influenced and inspired by many of his published works. Embracing a blend of empirical and theoretical positions, and recognizing the critical significance of the past in understanding present issues and in shaping policies for the future, the book provides a cohesive set of research statements on critical related issues in British rural geography, as well as echoing the priorities identified by an influential figure in British rural geography.

The first four chapters adopt a time perspective to explore a series of key themes: the distinctive character of rural geography in Britain, which is increasingly positioned in an interdisciplinary perspective; the rise of productivist farming; ways of conceptualizing agricultural change; and the evolution of landownership and property rights. Following these, a cluster of chapters emphasizes essential connections in rural geography: earth and life; nature and culture; rural and urban agendas for nature conservation; and the gap between policy and action for sustainable development. The final set of chapters is devoted to policy-related issues associated with agricultural change and the profound challenge of rural diversification for the future: tenant farmers and the CAP; and diversification of agricultural enterprises and rural areas. The last chapter traces the influential career of Richard Munton, one of Britain's most important rural and environmental geographers.

The book demonstrates that the rural world needs to be seen in a far wider perspective than that of agriculture and food production, in order to comprehend how resources are being appraised and exploited in new ways, and to respond to the pressing challenges of sustainability for the decades ahead.

Hugh Clout is Emeritus Professor of Geography at University College London where he introduced the teaching of rural geography in the late 1960s.

Contemporary Rural Geographies

Land, property and resources in Britain: essays in honour of Richard Munton

Edited by Hugh Clout

 Routledge
Taylor & Francis Group

LONDON AND NEW YORK

First published 2007
by Routledge
2 Park Square, Milton Park, Abingdon, Oxfordshire OX14 4RN

Simultaneously published in the USA and Canada
by Routledge
711 Third Avenue, New York, NY 10017

First issued in paperback 2014

Routledge is an imprint of the Taylor & Francis Group, an informa business

Typeset in Times by Wearset Ltd, Boldon, Tyne and Wear

British Library Cataloguing in Publication Data
A catalogue record for this book is available from the British Library

Library of Congress Cataloging in Publication Data
A catalog record for this book has been requested

ISBN13: 978-0-415-43183-5 (hbk)
ISBN13: 978-1-138-01064-2 (pbk)

Contents

Figures

Tables

Contributors

Jacquie Burgess moved to take up the chair in Environmental Risk in the University of East Anglia in autumn 2006 after 31 years in University College London's Department of Geography. Jacquie established the Environment and Society Research Unit (ESRU) in 1996, and collaborated with Richard Munton in developing ESRU's distinctive approach to sustainability governance.

Julian Clark is Lecturer in Environmental Governance at Birkbeck College, University of London, having completed his doctoral work at University College London (UCL) under the supervision of Alun Jones with whom he worked as Senior Research Fellow in the Geography Department at UCL. During 2002–3 he taught with Richard Munton on the M.Sc. Conservation course.

Paul Cloke lectured at the University of Wales (Lampeter) and was Professor of Geography in the University of Bristol, where Richard Munton was an influential external examiner. In 2005 Paul moved to the chair of Human Geography at the University of Exeter. He is founding editor of the *Journal of Rural Studies*, which was founded in 1985, and is a major international outlet for research in rural geography and related fields.

Hugh Clout is Emeritus Professor of Geography at UCL, where he taught in the Geography Department for 40 years. His several research interests include rural geography both past and present, and for two decades he delivered a pioneering undergraduate course in rural geography whose emphasis on social issues complemented Richard Munton's teaching in agricultural geography. Richard was head of department for two periods during the decade when Hugh was Dean of the Faculty of Social and Historical Sciences.

David Goode is Visiting Professor in the Department of Geography at UCL, having been Director of the London Ecology Unit from 1986 to 2000, and then Head of Environment in the Greater London Authority. Over the past 20 years he has had close links with Richard Munton in the development of strategic policies for the development of London's open spaces and sustainable development, a theme on which they have written jointly.

Carolyn Harrison is Emerita Professor of Geography at UCL, having spent her entire career at the college. As an active member of the Environment and Society Research Unit in the UCL Geography Department, she collaborated with Richard Munton on research projects throughout the 1980s and 1990s and on teaching programmes for the UCL M.Sc. Conservation Course.

Lewis Holloway is a Lecturer in Geography at the University of Hull, research-ing into 'reconnecting consumers, food and producers' as part of the Cultures of Consumption Programme. He researched his doctorate under the supervi-sion of Brian Ilbery and lectured at Coventry University from 1996 to 2004.

Brian Ilbery is Professor of Rural Studies in the Countryside and Community Research Unit, University of Gloucestershire. From 1977 to 2006 he taught at Coventry University, where he held a Chair of Human Geography. Brian has worked with Richard Munton for the Rural Geography Research Group of the Institute of British Geographers and has examined two of his doctoral students.

Alun Jones is Professor of Geography at University College Dublin. He was previ-ously Professor of Human Geography at the University of Leicester, and prior to that he taught at UCL, contributing to courses on rural geography and environ-mental policy at undergraduate and postgraduate levels with Richard Munton.

Philip Lowe is Duke of Northumberland Professor of Rural Economy at the University of Newcastle and director of the UK Research Councils' £25 million Rural Economy and Land Use Programme. Between 1974 and 1992 he was a Lecturer and Reader in Countryside Planning at the Bartlett School of Architecture and Planning at UCL and was a co-director of the Rural Studies Research Centre, with Richard Munton, from 1988 to 1992.

Terry Marsden is Professor of Environmental Policy and Planning in the School of City and Regional Planning at Cardiff University. He is a co-director of the Economic and Social Research Council's research centre on Business Relationships, Accountability, Sustainability and Society, also based at Cardiff. During the 1980s and early 1990s he worked intensively with Richard Munton in developing a revised theoretical and empirical understanding of the contemporary political economy of UK agriculture, and then as co-director (with Richard and with Philip Lowe) of the ESRC's Countryside Change Centre, based at UCL.

Damian Maye is a Lecturer in Human Geography at Coventry University. Brian Ilbery supervised his doctoral work and has worked with Damian on a number of research projects, including food-labelling schemes in Europe and North America; supply chain linkages for small food businesses in Europe's 'lagging' rural regions; and tenant farming and diversification.

Brian Short is Professor of Historical Geography at the University of Sussex, with a particular interest in the evaluation and use of twentieth-century archival material on rural Britain. A near-contemporary of Richard Munton, and with him an early member of the (then) Agricultural Geography Research Group of the Institute of British Geographers, his historical work has complemented the contemporary emphasis in Richard's research.

Neil Ward is Professor of Rural and Regional Development and Director of the Centre for Rural Economy at the University of Newcastle. He was taught by Richard Munton while an undergraduate at the Department of Geography, UCL, and worked as a Research Officer with Richard between 1988 and 1993. Richard also supervised Neil's Ph.D., which was awarded in 1994.

David Watts is a researcher at the Scottish Agricultural College in Aberdeen. He studied geography at the Universities of London and Bristol, and worked on a range of rural projects with Brian Ilbery at Coventry University from 2003 to 2006.

Sarah Whatmore is Professor of Environment and Public Policy at the University of Oxford and Director of the International Graduate School at Oxford University Centre for the Environment. Richard Munton supervised her Ph.D. at UCL (1985–8) and Sarah worked closely with him (and Terry Marsden) during the 1980s as a Research Associate on a series of ESRC projects concerned with agricultural change.

Michael Winter is Professor of Rural Policy and Director of the Centre for Rural Research at the University of Exeter. He was a founding member in 1978 of the Rural Economy and Society Study Group, and it was through the RESSG that he came to know Richard Munton in the early 1980s. He was an associate of the UCL Rural Studies Research Centre.

Preface

This book presents a collection of original research essays that cover major themes in the rural geography of contemporary Britain. They have been written in honour of Professor Richard Munton, who retired in September 2006 from the established chair of Geography at University College London. The chapters mirror components in Richard's research trajectory, their authors having been influenced and inspired by his personal support and by his many published works. Embracing a blend of empirical and theoretical investigation, and recognizing the critical significance of the past in understanding present issues and in shaping policies for the future, the chapters provide a set of statements on critical related issues in the rural geography of Britain. Varied in both scope and academic style, they exemplify a range of very different approaches to be found in the practice of the sub-discipline of rural geography, which is currently demonstrating environmental and cultural orientations as well as economic and social concerns.

The first group of chapters adopts a time perspective to explore four themes: the distinctive character of rural geography in Britain, which is set increasingly in an interdisciplinary perspective; the rise of productivist farming in wartime as an overlooked precursor to more recent intensification; innovative ways of conceptualizing agricultural change; and the evolution of landownership and property rights in recent decades. Philip Lowe and Neil Ward (Chapter 1) begin the volume by tracing the changing nature of rural geography in Britain, setting this academic enterprise in the wider academic context of changing debates in human geography and in related sub-disciplines such as agricultural economics, rural sociology and rural planning. They argue that British rural geography has been successful, in large part because of its openness and responsiveness to wider intellectual currents and public concerns. Nonetheless, the increasing emphasis in recent years on interdisciplinary scientific collaboration between the social and the natural sciences imposes new challenges for geographers who are interested in questions of rural land use and countryside change. Looking back to the middle of the twentieth century, Brian Short (Chapter 2) notes that while the concepts of productivism and post-productivism have been disputed as descriptors of recent agricultural change, there can be little doubt that between 1939 and 1960 many interacting stimuli drove British farming towards greater commercial and technical modernity. An essential component in this story of change was the establishment by the Ministry of Agriculture in September 1939 of the County War Agricultural Committees, charged with ensuring sufficiency of food in wartime. The chapter examines the structure and composition of those committees, their power, and their impact on the rural societies and environments of England and Wales.

After recalling earlier work written jointly with Richard Munton, Terry Marsden (Chapter 3) explores recent structural changes occurring in British agriculture. He argues that both theoretical and empirical approaches need to be embraced, together with an understanding of the social, economic and environmental implications of structural change. Now these concerns have more relevance than ever in the policy context of sustainable rural development, which has significant implications for the future size and structure of the farming sector. However, such considerations are largely left to the workings of 'the market'. Returning to another of Munton's early research themes, Michael Winter (Chapter 4) shows that British academics have paid little subsequent attention to landownership and property rights in recent decades. Yet in the late twentieth century there were major changes in agricultural holdings legislation as a result of the introduction of farm business tenancies in England and Wales, and a major land reform process in Scotland. Successive reforms of the Common Agricultural Policy and changes in both agricultural and land markets have contributed to a further growth of share farming and contract farming. Owners' rights to exclusive use of their land have been eroded by the extension of public access rights to some land. As well as appraising different approaches in rural geography and analysing essential trends in landownership and occupancy over the past 20 years, this chapter considers implications for environmental management and rural development.

The second group of chapters emphasizes essential connections within the broad scope of rural geography: between earth and life; nature–culture connections; nature conservation agendas that extend beyond the countryside to embrace urban environments; and the gap between policy and practice in terms of involving the public in action for sustainable development. Sarah Whatmore (Chapter 5) examines tensions between the territorializing and deterritorializing tendencies in the framing of property relations and then outlines the importance of new theoretical approaches to the understanding of property. Starting from work undertaken in the United Kingdom but then looking out to Australia, her chapter concentrates on the emergence of new forms of property, notably in association with innovations in the life sciences and with the proliferation of bioresources. It then explores how new conceptions of property throw light on everyday and conventional understandings of the ways in which property works. Also adopting a wider spatial framework than the United Kingdom alone, Paul Cloke (Chapter 6) argues that the differences between urban and rural are becoming blurred in the contemporary world, with all-pervasive processes of cultural urbanization and extended suburbanization serving to bring the urban out into the country and, to a lesser extent, with the attraction of rural ideals and style, bringing aspects of the country into the city. He maintains that perceptions of rural life and landscapes are being increasingly ordered and constructed by new forms of commodification, often associated with leisure and tourism, through which rural 'attractions' serve up sanitized and spectacular theatres of rurality. His chapter uses such changes as a basis for outlining a re-presentation of rurality through tourism. It traces examples of how tourists are being presented with opportunities for creative performativity in the countryside, and suggests that perceptions of rurality are being especially informed by these creative practices and performances. In some ways it would seem that the direct implication of nature–society relations in these new rural creativities is once again setting the countryside apart through motifs of adventure, eco-desirability and landscape heritage.

Operating in a very different style and drawing on his long experience as Director of the London Ecology Unit and latterly as Head of Environment at the Greater London Authority, David Goode (Chapter 7) reviews how schemes have been developed for nature conservation in British cities, a theme that is present in some of Richard Munton's most recent work. This chapter discusses the main differences of approach between urban and rural agendas for nature conservation and explores ways in which an ecological perspective can contribute to environmental sustainability in urban areas. Continuing with the theme of sustainability, Carolyn Harrison and Jacquie Burgess (Chapter 8) argue that public participation in environmental management in Britain has been slight, despite recent Labour governments' seeking to increase public involvement in public affairs through their devolution and modernization agenda. National agencies concerned with the environment have been encouraged to consult organized groups and individuals. In this reflexive chapter the authors review their research undertaken during the 1980s and 1990s, which experimented with various forms of dialogue and discursive space through which members of the public could engage with the process of land-use planning. Focusing on the protection and promotion of accessible green space, in both countryside and urban areas, they explore how national agencies and local publics responded to calls from government for public participation in environmental planning.

The final two chapters are devoted to policy-related issues associated with agricultural change and rural diversification, exploring the influence of one innovative aspect of the Common Agricultural Policy (CAP) on tenant farmers, and finally illuminating the mechanisms of diversification available to farm businesses. Brian Ilbery and his colleagues Damian Maye, Lewis Holloway and David Watts (Chapter 9) report on the findings of a research project on the potential impact of single farm payments (one of the major policy instruments of the 2003 reforms of the CAP) on the current and future diversification activities of tenant farmers in England. Information obtained by postal questionnaire surveys and by telephone interviews enables them to examine diversification activities in different types of agricultural area, including 'severely disadvantaged' locations. Julian Clark and Alun Jones (Chapter 10) further develop the theme of diversification of farm businesses in the United Kingdom, drawing upon detailed information from the English East Midlands. They trace the role of farmer networks in initiating and sustaining diversification objectives and then evaluate the impacts of diversification upon farm households, rural territories, and public policies for rural activities and environments.

In Chapter 11, I, as editor, chart Richard Munton's career as a researcher and teacher of rural geography, supervisor of many doctoral students on rural-environmental themes, adviser on landownership and other countryside topics to numerous government agencies, and member (and ultimately chair) of assessment panels evaluating the quality of geographical research in universities throughout the United Kingdom. This short biography is complemented by a bibliography of his publications (Appendix I). I wish to express my profound thanks to Malcolm Ward for invaluable assistance in assembling the chapters of this book entirely electronically.

Hugh Clout

1 British rural geography

A disciplinary enterprise in changing times

Philip Lowe and Neil Ward

Introduction: disciplines and sub-disciplines

Ron Johnston's studies of the evolution of human geography since 1945 remind us that the content of an academic discipline cannot be understood without reference to its context (Johnston 1991). The same can be said for sub-disciplines such as rural geography, which emerged in Britain in the 1970s. This chapter sets its history in the context of changing conceptual approaches and patterns of institutionalisation in human geography and other related fields, such as agricultural economics and rural sociology. It argues that British rural geography has been a successful sub-discipline of human geography, in large part because of its openness and responsiveness to wider intellectual currents and public concerns.

Most social scientific work is set within disciplines yet at the same time builds (or dismantles) those disciplines. The development of ideas, concepts and empirical inquiry is therefore inseparable from the act of constructing disciplines. Disciplines are structured contexts that are re-created through the efforts of scientists producing knowledge in pursuing their careers. Disciplinary structures provide exclusive access to research and career resources, and this is why boundary maintenance is so essential. Boundary maintenance occurs within a wider 'commonwealth' of scientific knowledge production in which boundaries between disciplines cannot legitimately be maintained simply through the exclusionary practices of a restricted trade or freemasonry but must be reproduced through recognised knowledge work. The epistemological structures of disciplines are therefore inevitably co-produced with their institutional structures, and reciprocally so with other disciplines.

A major part of disciplinary boundary work is the maintenance of sub-disciplines. Created typically at times of disciplinary expansion, sub-disciplines reflect the needs of scientists to differentiate what they produce and generate new professional niches. However, such a process may be forced into reverse. Cloke *et al.* (1991: 21) recount the salutary instance of American geography, a discipline for many years in retreat. They quote the President of the Association of American Geographers who warned, 'We are a small discipline. To be small these days is to be vulnerable', and his admonition that 'Geography's continual splitting into smaller clusters has become hazardous to our collective health'

(Abler 1987: 518). Similar forebodings were heard in UK geographical circles in the mid-1980s, but British geography proved more firmly rooted. Moreover, the outlook of young and ambitious researchers, even when the prospects for their discipline may seem unpromising, may not concur with the instincts of their seniors to consolidate. Munton and Goudie, for example, dismissed the 'unease expressed by some geographers about the continuing fissiparous tendencies within the discipline' (1984: 27), and the related calls to strengthen the traditional core areas of the subject, with the confident assertion that

> geographers will continue to draw upon theory derived from related disciplines in their search for explanation or greater understanding. The search for such theory may lead to centrifugal tendencies, but there need be no concern for the core if geographers take advantage of the numerous emerging areas of research to which many believe they can make a distinctive contribution.
>
> (ibid.: 39)

The dynamics of sub-disciplines thus illuminates not only centre–periphery relations in discipline building and knowledge production but also generational change within disciplines, as well as the competition between disciplines as they rub up together at the edges.

Andrew Abbott's book *Chaos of Disciplines* (2001) argues that disciplines evolve through essentially similar processes that he characterises as a fractal pattern of continuous internal division and occasional external convergence. Drawing on the example of the relationship between history and sociology, he concludes that supposedly contrasting disciplinary specialisms, such as social history and historical sociology, may have more in common with each other than with the mainstream of their parent disciplines, thus allowing considerable traffic of people and ideas between them. Abbott analyses how distinctions are played out over time and shows that when lines of inquiry wither away, their concerns are often subsumed by, or 'remapped' on to, other branches. Neighbouring disciplines and sub-disciplines therefore evolve through processes of engagement, dialogue, conflict, bifurcation and ingestion.

While Abbott's analysis focuses primarily on the internal dynamics of disciplinary development, he recognises that moments of differentiation and absorption can be shaped by external factors. These may be to do with newly recognised problems in the wider social or political world that may either enlarge or reduce the resources available for disciplinary expansion. Fractal processes expand to fill whatever space is available when disciplines expand. The emergence of environmental economics, environmental politics and environmental sociology in the 1980s and 1990s is an example. Conversely, less diversity and more concentration is the outcome when resources are more limited. However, the processes of coming together and breaking apart regularly occur and follow a similar pattern. One side loses and the winning side then becomes characterised by further fractal development and 'remapping' of the loser's interests and concerns.

Abbott's model provides a useful starting point from which to examine patterns and processes of innovation, consolidation and decline in the development of sub-disciplines, and we draw on it to analyse the way in which 'the rural' as a field of inquiry has been subsumed into the social sciences in the United Kingdom. We argue that the 1980s were a key decade in the emergence of the rural as a focus of social science inquiry. This was a period of considerable turmoil in the social sciences in the United Kingdom, marked by twin trends of radicalisation and professionalisation among academic social scientists. Besides other developments, the 1980s saw the birth of a critical interdisciplinary rural studies that then, and subsequently, has provided fertile ground for interchange between disciplines, re-energising some but depleting others. Rural geography, in particular, was able to grow in strength at the expense of other sub-disciplines such as rural sociology and agricultural economics. In analysing the dynamic interaction of the disciplines, we start first with a brief review of the status of rurally oriented social sciences up to the 1970s.

Rural studies in the United Kingdom up to the 1970s

The constitution of the rural world and the dynamics of agrarian change were major preoccupations of the classic social science disciplines: economics, geography and history. Only in economics had this been institutionalised as a distinct sub-discipline in Britain, that of agricultural economics. However, rural topics pervaded mainstream scholarship in history and geography, with considerable overlap in interests in the parallel sub-disciplines of agrarian history and historical geography.

Agricultural history had always played an important part in the syllabi of British economic and social history degrees. Questions of agricultural change, land use and settlement systems were even more central to the discipline of geography, however. Indeed, the rural landscape was key to the main traditions of geographical research, of regional survey and historical geography. The epithet 'rural geography' would have smacked at least a little of tautology. It is notable, for example, that the prominent geographer David Harvey has never been identified as a 'rural geographer' (Castree and Gregory 2006), even though his doctoral research, completed at Cambridge in 1962, and early publications examined agricultural and land-use change in rural Kent (Harvey 1962, 1963, 1966, 2006). Harvey, of course, went on to be a leading figure, in quick succession, in spatial science and then Marxist urban geography.

The trajectory of one of his contemporaries, Ray Pahl, is also illuminating. Having been associated as a postgraduate with the advent of the 'new geography' (namely spatial science), his Ph.D. which was the first study of the impact of counter-urbanisation on a village community, would surely qualify him as a father figure of either rural geography or rural sociology (Pahl 1965a, b). However, he moved quickly into the sociology of urban planning (developing critiques not dissimilar from Harvey's) and was soon a professor of sociology.

If the rural was not a destination for avant-garde geographers by the 1970s, there was nevertheless a growing demand from geography students for teaching that covered rural issues beyond the specific realm of agricultural production, whose geographical researchers were headed by J. Terry Coppock (Clout 2002). Hugh Clout produced the first rural geography text in 1972 (Clout 1972). Recently, he has reflected that this was done 'unashamedly to plug a gap in the textbook literature of the time' (Clout 2005: 376). Most revealingly, he goes on:

> My cultivation of 'rural geography' at University College London was part of a personal survival strategy. I simply felt that I required another systematic support to complement my main interests in historical geography and France. The textbook – along with several others – was woven into a 'personal safety blanket' to keep me going since my part-time doctoral research advanced slowly.[1]

Geography was still a strongly teaching-oriented discipline (Stoddart 1986). The production of textbooks, then and now, explicated fields of study for an undergraduate audience while also demarcating areas of scholarly competence (see, for example, Phillips and Williams 1984; Pacione 1984; Gilg 1985; Woods 2005).

The Geography Department at University College London in the early 1970s was stocked with historical geographers (Clout 2003). Once the foundation of British geography, historical geography was undergoing something of an identity crisis as the post-war American vogue for spatial science swept through the discipline. For a young lecturer trained in historical geography, a focus on contemporary rural issues provided a new distinctive niche. As Clout explained:

> My *Rural Geography* omitted any reference to the 'less developed world' and excluded all the historical material that I taught as rural geography to large numbers of second year undergraduates at UCL in the late 1960s and during the 1970s.
>
> (2005: 376)

With the swelling number of undergraduate geographers in the post-Robbins expansion of the universities, there was a need to give geography teaching a more contemporary appeal. It was this requirement rather than academic fashion that drove the delivery of rural geography. As Clout himself observed at the time, with academic and professional attention increasingly focusing on the quantitative analysis of urban and regional change, the rural had been relegated from being at the core of geography to an inferior position. His proposed solution was to refocus academic skills on the problems of relevance to countryside management and rural planning. The renaming of the Agricultural Geography Study Group of the Institute of British Geographers (IBG), founded by Coppock, as the Rural Geography Study Group in 1974 should thus be seen as an effort to refurbish a field of activity that was in danger of becoming marginalised within contemporary geographical research. The focus shifted to the geographical analysis of rural problems and away from efforts to delineate the agricultural regions of various parts of the world. The Rural Geography Study Group was one of the Institute's most popular. By 1983 its membership stood at 280, which placed it fourth among the IBG's study groups, after 'urban', 'quantitative' and 'geomorphology'.

Agricultural economics was the only rural sub-discipline that was institutionalised in the United Kingdom, with agricultural economics departments in the Universities of London (at Wye), Reading, Oxford, Exeter, Cambridge, Nottingham, Manchester, Newcastle, Edinburgh and Aberdeen. These departments had their roots in the early 1920s, when the Ministry of Agriculture had established provincial advisory centres in universities and colleges to provide advice to farmers. Each of these centres was equipped with an agricultural economist, and they built up expertise in farm management and production economics, although some also pursued other interests, including rural sociology and history (see Colman 1990). Agricultural economics also became a recognised specialism within government, and economists in the Ministry of Agriculture maintained a close professional relationship with agricultural economists in the universities and colleges. The Agricultural Economics Society, founded in 1926, brought the two groups together, and with its *Journal of Agricultural Economics* (begun in 1928) the society became an influential forum for the analysis of the state of agriculture and agricultural policy. After the Second World War, the government reorganised what had been the provincial advisory centres into a separate government research and extension service, but agricultural economics was left with the universities and colleges. With a much-reduced advisory role, agricultural economists concentrated on the development of teaching and research. In some cases they were brought together with academic economics departments, but in most cases they existed in separate agricultural economics departments, often alongside departments of agriculture. Many of the younger agricultural economists in post in the 1960s and 1970s did graduate training in American universities, where they were subjected to a more rigorous theoretical and mathematical training than had been available in the United Kingdom. Back home, they re-established agricultural economics on a stronger basis of neo-classical welfare and trade theory and, in particular, a thoroughgoing and highly quantitative pursuit of inferential econometric methods. As a cohesive and well-institutionalised discipline, agricultural economics thus dominated social science research on agricultural issues throughout the post-war period.

In the United States and most other European countries, an institutionalised rural sociology sat alongside agricultural economics, because the post-war modernisation of agriculture was seen to imply a wider transformation of rural society that went beyond improving the productivity of farm labour. No such institutionalised rural sociology, dedicated to easing the pace and effects of agrarian transformation, existed in the United Kingdom. Among sociologists and social anthropologists interested in kinship there was, though, a tradition of 'localistic studies' that in the 1950s and 1960s had examined the cohesiveness of isolated farming communities, but whose findings had served to undermine the very assumptions on which they had been based concerning the intrinsic and abiding characteristics of rural communities (Bell and Newby 1971). A range of British scholars, including many who were not sociologists but who took an interest in the social aspects of farming or village life, belonged to the European Society for Rural Sociology, founded in 1957, and contributed to its journal *Sociologia Ruralis*.

The predominance of agricultural economics in the United Kingdom left little scope for the institutional development of rural sociology. Often through American graduate school experience, many UK agricultural economists had been exposed to rural sociological ideas, and some showed a professional interest in social analysis of farming and rural communities (see, for example, Wibberley 1960; Gasson 1971; Jones 1973). Agricultural economics departments, however, were accused of 'exclusionary practices' by the sociologist Peter Hamilton, in not appointing professional sociologists and in allowing issues that might have been the subject matter of rural sociologists (such as agricultural labour mobility or farm management) to be 'hived off into strange culs-de-sac by agricultural economists and given obfuscatory names such as "agricultural adjustment"' (1990: 229). More generally, he lamented the fact that there had been no demand in post-war Britain for a broader rural sociology, 'either from the agricultural sector or rural society, but more significantly ... from the profession of sociology itself' (ibid.: 229). This was a function of the fact that sociology as a discipline had only a limited intellectual purchase in Britain and, seen as a science for understanding and resolving social problems, it was preoccupied with urban and industrial issues.

The 1980s: the birth of interdisciplinary rural studies

The 1980s were a pivotal decade in the evolution of rural studies and more generally as critical social theory swept across the social sciences. This was a period of a polarised and ideologically charged politics following the breakdown of the dominant post-war social consensus. The empiricism and positivism that had marked most British social science fell out of favour. A new generation of young researchers radicalised by the student politics of the late 1960s looked to professionalise their work. However, the lack of secure career prospects within academic departments detached them from allegiance to established disciplinary perspectives and traditions. In response to the attacks from the New Right, they sought to institute a notion of independent academic study as a critical conscience within society, and found inspiration in European structuralist and post-structuralist theorists. One of the thrusts of critical social theory was to challenge academic conventions, which gave licence to those wanting to transcend academic boundaries.

In the rural sphere, interdisciplinary dialogue between the social sciences was facilitated by the activities of the Rural Economy and Society Study Group (RESSG). Established in 1979 as 'a forum for all those studying the social formation of rural areas in advanced societies [and] to encourage theoretically informed investigation and analysis of rural issues' (Bradley and Lowe 1984: 1), the RESSG brought together isolated rural scholars and previously diffuse networks from across the social sciences. It soon grew to include more than 100, mainly younger, researchers and went on regularly to attract research council funding for organising major conferences on rural themes. Its work can be traced through a series of edited volumes (see, for example, Bradley and Lowe 1984; Cox *et al.* 1986; Lowe *et al.* 1987; Bouquet and Winter 1987; Buller and Wright 1990; Marsden and Little 1990; Milbourne 1997).

A major outlet for interdisciplinary rural studies in the United Kingdom has been the *Journal of Rural Studies*, edited by the geographer Paul Cloke since its launch in 1985. The journal has become a focal point for the publication of rural studies research, across a range of disciplines, and has a strongly international reach. In his opening editorial, Cloke reflected on how rural geographers had been drawn out of a 'fallow period' by 'a direct and compelling exposure to other perspectives on rural areas, which has widened their horizons way beyond the bounds of traditional geographic study' (1985: 2).

The first wave of theoretical development drew upon concepts and ideas from political economy, such as the centrality of capital accumulation and restructuring in social formation and uneven development. Howard Newby's 1970s studies of the social relations of capitalist farming were particularly influential. They broke from the tradition of rural community studies and established theoretical interest in the power relations between rural social groups. Starting with his doctoral work on the deferential behaviour of farmworkers, Newby's researches opened out into a wide-ranging analysis of the changing social structure of lowland Britain, encompassing the social and economic status of farmworkers, small farmers, 'agribusiness' farmers and the ex-urban middle class, based on an analysis of rural property relationships and the rationalising tendencies of a state-sponsored capital-intensive agricultural industry (Newby 1977; Newby *et al.* 1978). This encompassing analysis, drawing on and exemplifying general and ascendant sociological concepts such as class and power, transformed the standing of British rural sociology, or perhaps, more precisely, the sociology of rural Britain. More broadly, in embracing new issues such as the social and environmental impacts of agribusinesses and the urbanisation of rural social structures, Newby's work demonstrated how the study of rural change could illustrate and illuminate general processes of social and economic change in contemporary Britain (Newby 1980). Freed from a sense that 'the rural' was an archaic backwater isolated from the contemporary world, rural studies became fashionable, led by Newby's self-proclaimed 'New' or 'Critical Rural Sociology' (Newby and Buttel 1980; Newby 1983).

The impulse of the new rural sociology, however, was felt more strongly outside British sociology than within it. Noticeably, none of the leading sociologists, such as Ray Pahl, John Urry, Ian Carter and Howard Newby himself, who so strongly influenced social science understandings of contemporary rural Britain, spun off 'schools' of rural sociology among their graduates. Looking back in 1990, Graham Crow, one of Newby's former postgraduates, commented: 'Since Newby's research of the 1970s, few sociological farm studies of any sort have been undertaken, reflecting the fact that British sociology continues to have an overwhelming urban orientation' (Crow *et al.* 1990: 253). To Peter Hamilton, the failure of rural sociology to achieve an academic foothold was due to the fact that 'the lack of institutional locations at a time when the university sector in the UK, and sociology in particular, are contracting has proved a formidable handicap to the formation of a coherent grouping' (1990: 230).

Of course, institutionalised rural sociology existed outside the United Kingdom, and some of the pioneers of critical rural sociology did find chairs abroad in the 1980s, for example Ian Carter in Australia and Howard Newby in the United States. In the United Kingdom, others picked up their ideas. As Crow *et al.* commented:

> [B]ecause of the paucity of institutional support and sponsorship for rural sociology in Britain, much work which is clearly influenced by ideas from sociology is produced within other disciplines with a strong tradition of rural studies, for example rural and social geography, planning and social anthropology.
>
> (1990: 254)

Mainstream sociological attention moved on, and indeed was encouraged to do so by the very figures who had trail-blazed the sociology of rural Britain and who, in dismissing the notion of a distinctively rural economy and society, had incidentally also deconstructed the rural as a sociological category. Pahl had concluded from his study of commuter villages that 'in the sociological context, the terms rural and urban are more remarkable for their ability to confuse than for their power to illuminate' (1966: 299) and argued that 'any attempt to tie particular patterns of social relationships to specific geographical milieux is a singularly fruitless exercise' (ibid.: 322). For his part, John Urry, having characterised the rural as an historically contingent and descriptive category which lacked explanatory power, then dismissed it as a 'chaotic conception' (Urry 1984). Newby too had berated an ultimately 'futile search for a sociological definition of "rural"', which he ascribed to 'a reluctance to recognise that the term "rural" is an empirical category rather than a sociological one ... it is merely a "geographical expression"' (Newby and Buttel 1980: 4). Significantly, though, by the early 1980s, Newby was addressing an international, mainly American, *rural* sociological audience and he indicated to them a way forward out of the theoretical impasse:

> [I]t is not only apparent that rural sociology cannot operate without an acceptable theory of society, but that it also requires a theory of the spatial allocation of the population (since 'rural' is a spatial, geographical category) which is also sociologically relevant. In other words, rural sociology demands a theory that links the spatial with the social.
>
> (ibid.: 5–6)

Newby recognised this as 'fruitful – if contested – theoretical and empirical terrain' (ibid.: 27) but saw it as the way to build a comparative rural sociology of advanced capitalism.

Compared with other social science disciplines, such as linguistics, economic and social history, anthropology and sociology, critical social theory came late to geography, and in many cases came via these other disciplines. This lag effect reflected the strength of positivist and empiricist traditions in geography. However, the vogue for spatial science had run its course and was encountering mounting criticisms not only of a methodological but also of an ontological kind. A counter-movement developed to promote 'post-positivist' approaches to human geography, and this systematically reconstituted the discipline through the 1980s and 1990s as a self-consciously theoretically oriented social science. In the words of one participant, 'theoretical sophistication has replaced quantitative/analytical sophistication as the guarantor of geography's legitimacy' (Robert Lake, quoted by Cloke *et al.* 1991: 207). The transformation of geography from a discipline with little if any tradition of reflexive epistemology to one with a 'high degree of theoretical plurality' (Hubbard *et al.* 2002: 6) involved an eclectic borrowing of ideas and methods from other social sciences, as described by Hubbard and colleagues:

> Often, [geographical] theorizing involves adapting theories developed in other disciplines, such as social theory, economic theory or political theory. Ultimately, this 'borrowing' of ideas from other disciplines serves to make the strange familiar, and helps geographers to conceive of the world in new ways.
>
> (ibid.: 4)

This is not the place to document or account for the rise of human geography to be such a prominent social science in Britain in the 1990s. Central to any fundamental explanation would be the contradictory position of the social sciences in the United Kingdom in the 1980s: on the one hand, facing ideological opposition and funding cuts from the government; on the other hand, under pressure to expand to fulfil society's continuing requirements for social engineering and to support the mass expansion of higher education. While other social sciences, except for economics, business studies and management, were actively disfavoured, geography in the United Kingdom was able to weather the challenge of the 'New Right', thanks to its entrenched position in the UK educational system and its apolitical public image.

Writing in 1984, Munton and Goudie captured the sense of insecurity and uncertainty but also of intellectual opportunity felt by UK geographers at the time. They recorded how government funding cuts in the early 1980s had marked an end to geography's post-war growth, but that neither the range of research activity nor the level of publication had diminished. However, human geography was in the grips of an 'unresolved epistemological debate' (Munton and Goudie 1984: 27), subjected to wide-ranging structuralist and humanist critiques, but with much of the profession resisting the implications of the debate:

> Although the critiques of positivism are widely acknowledged throughout the literature of human geography and the roles of values and ideology, both implicit and explicit, in the conduct of research are generally recognized, acknowledgement of these changes should not be equated either with their endorsement or with their ready translation into programmes of research.
>
> (ibid.: 34)

Noting in passing that rural geography was one of the last redoubts of an unreconstructed positivism, Munton and Goudie warned nevertheless that wider epistemological developments had 'attenuated the traditional content of sub-areas of the discipline' (ibid.: 30) and ineluctably were 'drawing rural and urban geographers into more general discussions on macroeconomic and social processes' (ibid.: 35). The following year, Richard Munton began his influential series of studies of the political economy of British agriculture, conducted with Terry Marsden, Sarah Whatmore and others, that revealed the different degrees of capitalist subsumption of family farming in different parts of rural England (Marsden *et al.* 1986, 1987; Whatmore *et al.* 1987; Clout, this volume, Chapter 11). The work was recognised internationally as showing how it was possible to explain the specificity of agrarian structures in time and place and thus to move theoretical debate beyond an overly deterministic structuralism (Buttel *et al.* 1990: 173).

With geographers increasingly looking beyond their discipline for theoretical inspiration, they turned to various sources. Rural geographers could draw on rural sociology and other areas of rural studies in the United Kingdom and elsewhere, and for a period these became important avenues for critical social theory to enter geography, through such channels as the RESSG and the European Society for Rural Sociology. British rural studies was host in the 1980s to lively debates on the relationship between agricultural organisation and rural social structure, the diversity of agricultural production forms, the recomposition of rural society in the face of capitalist restructuring, and the courses and consequences of rural deprivation, informed by Newby's work and various structuralist theoretical frameworks.[2] At the same time, the revival of interest in the ethnography of rural Britain was informed by different schools of postmodern and interpretive thought on the construction of identities (Cohen 1990). For its part, North American rural sociology in the 1980s was taken up with the political economy of agriculture and the historical causes of uneven capitalist development, as reflected in agricultural structures and technological developments, and informed by classical Marxism (Buttel *et al.* 1990). Finally, European rural sociology was interested in the differentiation of rural social structures, processes of rural development, the governance of agrarian change and the regulation of food chains, informed by structuralist and post-structuralist theories (Bodiguel and Hervieu 1990).

Ideas and approaches subsequently absorbed into mainstream human geography and as diverse as regulation theory, international regimes, actor–network theory and ethnographic method found application first in rural contexts in other disciplines. Through the avenue of interdisciplinary rural studies, rural geographers were exposed early to these diverse traditions. This not only revitalised rural geography but also changed its standing within geography. Of all the rural sub-disciplines, therefore, rural geography was most influenced by interdisciplinary rural studies.

Critical social theory poured into human geography in the late 1980s through many other channels, but interdisciplinary rural studies was a crucial bridgehead specifically for rural geographers. There are various ways to chart how the sub-discipline was reoriented consequentially. The journal *Progress in Human Geography* is and was one of the main means of discipline-building, and through the mid-1980s Ian Bowler contributed the annual review of agricultural geography. His first review, for 1984, was divided into two sections: the shorter devoted to the work of 'a minority of researchers [who] continue to explore traditional themes in agricultural geography, although often with the appearance of working under conditions of diminishing marginal returns'; and the longer section devoted to 'a number of agricultural "issues" which attract interdisciplinary attention', including an increasing proportion of geographers, but for which 'the distinction between geographical and non-geographical research becomes arbitrary' (Bowler 1984: 255). Bowler referred to 'fruitful exchanges with rural sociologists' but looked forward to building stronger links with industrial economists to strengthen understanding of developments 'upstream' and 'downstream' of the agricultural sector (ibid.: 259). Traditionally, the external influences on agricultural geography had come mainly from agricultural economics, and in his 1986 review Bowler included 21 references from economics or agricultural economics, three times as many as the seven from sociology. The following year, however, the situation was reversed, with there being more than twice as many rural sociological references in his bibliography as economics ones, and Bowler opened his report with the observation that 'Marxist political economy, either implicitly or explicitly, underpins much of the recent literature on the processes of change operating in agriculture' (1987: 425). The next year, 1988, he opened his review with the following wry reflection:

> For some years the subfields of agricultural geography and agricultural economics have appeared to lie on converging paths of development. Increasingly, their subject matters and modes of analysis have been interchanged. … But the situation seems set to change with the growing application of political economy approaches in agricultural geography.
>
> (1988: 538)

With Brian Ilbery he set out a programme in the journal *Area* to overcome the myopic nature of teaching and research in agricultural geography and to provide it with new stimuli by refounding it to embrace political economy perspectives, adopt the organising concept of the food chain and analyse agriculture within the context of the wider rural economy (Bowler and Ilbery 1987). Terry Marsden responded with a note of caution to Bowler and Ilbery's manifesto that he deemed 'diversionary' (Marsden 1988: 315). He argued that the adoption of political economy approaches had not so much increased the coherence of the sub-field of agricultural geography as broken down disciplinary boundaries and encouraged greater dialogue with the work of non-geographers, particularly rural sociologists. He specifically rejected any attempt at disciplinary closure, arguing instead for 'an interdisciplinary effort whereby the boundaries of sub-disciplines are progressively weakened' (ibid.: 320). There was evidently more at stake than giving British rural geography a facelift, including the question of who rightly could claim the mantle of the 'new rural sociology' and what their international standing would be. Sarah Whatmore, a research colleague of Marsden, took over from Bowler in producing the annual agricultural geography reports for *Progress in Human Geography*. Indicating the shape of things to come, she looked to the influence not only of political economy perspectives but also of international 'poststructuralist perspectives and feminist scholarship ... filtering through from anthropology, rural sociology and other fields to geography' (Whatmore 1991: 308).

Symptomatic of this intellectual maelstrom was a continuing, and sometimes agonised, debate about the epistemological status of the rural. The rise of political economy approaches had raised questions about the distinctiveness of rural societies and economies, and, as we have seen, sociologists had dismissed the 'rural' as an interpretive concept. To set rural sociology back on track, Newby and Buttel had argued, 'demands a theory which links the spatial with the social' (1980: 6). In responding to this challenge, Bradley and Lowe (1984: 7) argued for a return to the question of 'the social meaning of local diversity in the advanced societies' and suggested that the key to this would be a combined understanding of 'the specificity of local social processes and the social distinctiveness of localities', essentially drawing together the political economy of localities and the ethnography of localism, operationalised through a comparative study of localities (ibid.: 5). They accepted that the notion of the rural had 'no heuristic value' (ibid.: 13).

Such a perspective posed a particular challenge for traditional rural geography. In his opening *Journal of Rural Studies* editorial, Paul Cloke had echoed Richard Munton's warnings about 'aspatial' social theory. He argued that 'rural studies as a framework of study may be threatened if social science continues to espouse structuralist epistemologies with their aspatial connotations' (Cloke 1985: 1). Later, he wrote of how

> such axioms as social relations being as important as spatial differentiation, and the category 'rural' being of low importance as an explanatory device [were] potentially destructive to the institution of rural geography – an institution which had been carefully nurtured into an important position within geography.
>
> (Cloke, in Cloke and Moseley 1990: 125)

Cloke thus diagnosed 'the issue of rurality' as 'the focus of a perpetual identity crisis for rural geographers', who faced a 'major conceptual hurdle ... to place rurality in the context of critical social theory' (ibid.: 128). After all, their professional niche has been predicated on an ability to exploit the explanatory power of rural characteristics, to point up differences about rural environments, to promote rural courses within geography and to sustain a flow of rural literature (ibid.: 128).

The 1990s: redefining rurality and refounding rural geography

The discipline least affected by the rise of critical social theory and interdisciplinary rural studies was agricultural economics. During the 1980s the New Right attack on the social sciences had exempted economics, which allowed it to assume an even more unrivalled status in government and academia than it had previously occupied. To a certain extent, agricultural economics could bask in a reflected glory. However, the Thatcherite approach to the support of university science was to emphasise basic and fundamental research rather than applied and 'near-market' research. At the same time, Thatcherism sought to roll back the state, and New Right politicians became increasingly critical of state intervention in various economic sectors, including agriculture. Agricultural economists themselves contributed powerful critiques of agricultural policy, but in supporting free-market reforms they also undermined the very rationale for their own subject, which had come into being to underpin the state management of agriculture. The response of the sub-discipline staved off any sense of identity crisis by reinforcing its identification with pure economics.

Agricultural economists were therefore even more anxious to distance themselves from the practical and commercial orientation of their farm advisory predecessors that had attracted the disdain of fellow academics. Those interested, for example, in farm management, agricultural marketing or farm extension were increasingly marginalised, and set up their own specialist groups separate from the Agricultural Economics Society. Leading agricultural economists were keen to identify themselves as, or at least with, academic economists. This led to a strict emphasis on rigorous and abstract quantitative analysis, applying the tools and concepts of mainstream economics, particularly microeconomics, neo-classical welfare theory and econometrics, to the agricultural sector. Colman explained that to succeed,

> aspiring economists of agriculture have to have an aptitude for abstract and quantitative analysis in order to be able to communicate to general economists and statisticians and to select the appropriate tools of analysis to apply to the agricultural sector.
>
> (1990: 170)

The 1990s saw a profound decline in agricultural economics in the United Kingdom. The reasons for this decline would warrant a study. Certainly, the political economy of university teaching was an important factor, with the shrinking economic, social and political role of UK agriculture. With fewer and fewer students choosing to study agricultural economics, degree programmes were closed down and departments were merged or disbanded. Undergraduate degree programmes in the subject are now practically absent from the British scene.

Neo-classical economic theory gave a strong theoretical underpinning to British agricultural economics, and while this provided a powerful set of tools for working within disciplinary conventions, it also contributed to what Colman (1990: 174) called 'a degree of insularity from some other disciplines'. The increasing interest in environmental issues prompted some limited diversification of research questions, but also offered younger agricultural economists a chance to escape to the more fashionable fields of environmental or resource economics. Intellectually more radical calls for a conceptual repositioning of the discipline with respect to wider currents in social and economic theory fell on deaf ears (Midmore 1996).

The difficulty with their self-defined secondary status is that agricultural economists regarded themselves as irredeemably second-class citizens to mainstream economists. Indeed, they had a preference for appointing young theoretically oriented and quantitatively skilled pure economists without any background in agriculture, to 'maintain and refresh the academic standing of the profession in the eyes of colleagues in related disciplines', namely economics and statistics (Colman 1990: 170). They also shunned options to diversify that might seem too tarnished by their practical or policy orientation. Gradually, it seemed, they ran out of respectable topics to study. As Ken Thomson commented in his 2001 Presidential Address to the Agricultural Economics Society:

> The decline in agricultural economics as a separate academic sub-discipline in Britain has been considerable in recent years. There have of course remained many problems in the core area of farm business management, agricultural commodity trade, and agricultural policy. However, these problems have become so complex or routine (or both) that they have been left largely to the tender mercies of government economists or private consultants.

(2001: 8)

He identified the lack of a national institute of agricultural economics as another reason for the subject's weakness and vulnerability, but also pointed to the reluctance of agricultural economists to engage with new policy debates, such as those around rural development. In going on to propose rural economic development as a fresh focus, he himself seemed to provide the kiss of death in identifying 'economic geographers as our main colleagues or competitors [with] mainstream economists usually being rather spectacularly uninterested' (ibid.: 8). With an air of decline and dejection pervading the discipline, agricultural economists have been unable to defend their professional territory. Several established chairs either have ceased to exist or have been reallocated to non-economists. The withering of this previously dominant sub-discipline has thus opened up new territory for colonisation by other disciplines.

In contrast, rural geography in Britain experienced a renaissance in the 1990s. Expanding research activity and scientific horizons increasingly defined the sub-discipline rather than teaching and textbook authorship. Paul Cloke in a *Journal of Rural Studies* editorial in 1997 captured the sense of expansion and optimism. He wrote: 'Rural geography, and rural studies more generally, seems to be undergoing something of a resurgence in Britain at the moment' (1997: 367). The number of papers submitted to the journal was rising. The ranks of 'rural geographers' were expanding. There was a 'boom in excellent and challenging graduate research studies associated with the rural' (ibid.: 371). Above all, there was 'a resounding shake-off of any rural inferiority complex, as some of the brightest (and often younger) human geographers and other social scientists have taken a theoretical interest in rural societies and spaces' (ibid.: 371).

An important, indeed crucial, development was the reclaiming of rurality as an appropriate scientific object. This came in large part through an influential exchange between Chris Philo (1992, 1993) and Jonathan Murdoch and Andy Pratt (1993, 1994) about the implications of postmodernism for rural studies in general and rural geography in particular. Philo called for greater consideration in rural geography of marginalised groups and communities, introducing the so-called cultural turn. Murdoch and Pratt's response was to formulate a broader sociological analysis of rural studies that incorporated the social construction of 'the rural' (see also Halfacree 1993). Their argument was in keeping with Massey's line that although space may be a social construct, social relations are clearly constructed in and through space, and therefore to ignore space is to ignore the arena of social construction (see Massey 1985). This rescuing of the rural as an object of analysis opened up a rich seam of new research on representations of rurality, rural identities and processes of social and cultural marginalisation (see, for example, Milbourne 1997; Cloke and Little 1997).

Conclusions: a disciplinary enterprise in changing times

Rural geography in Britain has thrived over the past two decades, in part due to its openness to interdisciplinary influences. Its fortunes contrast starkly with those of agricultural economics, which has declined largely because it was closed to such influences. The question arises as to how one sub-discipline could be so much more flexible and open than the other. Here we draw together our reflections and conclusions around this question, comparing the experience of the obvious winner and loser in the competition between the sub-disciplines we have described.

Viewed externally, the changing fortunes of rural geography and agricultural economics present a paradox. Agricultural economics began the Thatcher period as a very strongly institutionalised sub-discipline, with its own departments. Dominated by a single strong and confident paradigm, that of neo-classical economics, it had a clear sense of purpose and intellectual direction. In contrast, rural geography as a sub-discipline of human geography had little institutional expression beyond a study group of the IBG, and lacked intellectual dynamism. Rural geographers could not even be sure of the epistemological status of the defining object of their analysis: the rural. Moreover, the political climate of the 1980s was much more favourable to agricultural economics, with the New Right attack on the social sciences exempting economics. Agriculture may have been a shrinking sector of the economy, but there was no shortage of other topical research agendas that agricultural economics was well placed to pursue, including environmental economics and the economics of rural development.

To explain the changing fortunes of these two sub-disciplines through the 1980s and subsequently, we must therefore look to internal factors, including the relationship between each sub-discipline and its parent. Here, it seems, the separate institutionalisation of agricultural economics served to limit its flexibility and presented an obstacle to sub-disciplinary renewal by segregating it from mainstream economics, just as it was isolated from other social science disciplines. Rural geography, on the other hand, benefited from not being an institutionalised sub-discipline while being and remaining fairly central to its parent discipline. That continuously posed the issue of what or who constituted rural geography. While the former – the what – caused much anguish at times, the latter – the who – was basically a matter of assumed or attributed identity. Becoming or being a 'rural geographer' is largely a question of self-identification (and might be one of a number of overlapping identities). As a consequence, rural geography could be much more fluid and flexible not only in its exchanges with the parent discipline but with neighbouring sub-disciplines too.

With loose kinship rules, rural geography was open to the flux of people and ideas in critical interdisciplinary rural studies. Agricultural economics, on the other hand, had tight kinship rules based on econometric competence, rules that perversely were negatively disposed towards agricultural economists themselves (in deference to pure economists). Agricultural economics was seen by pure economists as a poor relation, and indeed felt itself to be so. Even the most eminent academics within agricultural economics had little profile in the main discipline. In contrast, leading rural geographers could be prominent figures in human geography.

Given its loose kinship rules, it was imperative for rural geography to have a clear and defensible epistemological status. Rural geography thus had a continuing preoccupation with what constitutes the rural, achieving a certain settlement on this question (for the time being, at least) in the mid-1990s. As a separately institutionalised sub-discipline, agricultural economics exhibited no equivalent compulsion towards epistemological self-reflection. Indeed, its preoccupation with applying theory and methods from mainstream economics led it systematically and routinely to downplay the distinctiveness of its subject matter: agriculture. As agricultural economics has 'given up the ghost', so rural geography has been able successfully to redefine its terrain of activity to transcend the problem of the declining social and economic importance of farming.

Notes

1 In fact, Clout's subsequent research career focused on the recent historical geography of France, with rural geography occupying a background position. The doctoral research eventually yielded two distinct theses (and two books), one on French agriculture *c.* 1840, for the University of London, and the other on rural land-use change in nineteenth-century France, for the Sorbonne. An earlier doctoral project on countryside change in the Auvergne yielded articles but no thesis (Clout forthcoming).

2 Reviewing the RESSG's first volume, *Locality and Rurality* (Bradley and Lowe 1984), the American rural sociologist Fred Buttel (1985: 190) commented:

> [O]ne can observe that in all of the articles, save two,... reference is made to the work of Howard Newby, whom many outside of the UK see as having been synonymous with UK rural sociology over the past decade. But there is only a scattered representation of Newby's major theoretical ideas in this volume (nor an article by Newby himself). I am still puzzling over whether this volume can be seen as the beginning of either a neo-Newby or post-Newby phase of British rural sociology.

References

Abbott, A. (2001) *Chaos of Disciplines*, Chicago: University of Chicago Press.

Abler, R. (1987) 'What shall we say? To whom shall we speak?', *Annals of the Association of American Geographers*, vol. 77, 511–24.

Bell, C. and Newby, H. (1971) *Community Studies*, London: George Allen & Unwin.

Bodiguel, M. and Hervieu, B. (1990) 'The metamorphosis of French rural sociology', in P. Lowe and M. Bodiguel (eds) *Rural Studies in Britain and France*, London: Belhaven, pp. 233–47.

Bouquet, M. and Winter, M. (eds) (1987) *Who from Their Labours Rest: Conflict and Practice in Rural Tourism*, Aldershot: Avebury.

Bowler, I. (1984) 'Agricultural geography', *Progress in Human Geography*, vol. 8, 255–62.

Bowler, I. (1986) 'Agricultural geography', *Progress in Human Geography*, vol. 10, 249–57.

Bowler, I. (1987) 'Agricultural geography', *Progress in Human Geography*, vol. 11, 425–32.

Bowler, I. (1988) 'Agricultural geography', *Progress in Human Geography*, vol. 12, 538–48.

Bowler, I. and Ilbery, B. (1987) 'Redefining agricultural geography', *Area*, vol. 19, 327–32.

Bradley, T. and Lowe, P. (eds) (1984) *Locality and Rurality: Economy and Society in Rural Regions*, Norwich: GeoBooks.

Buller, H. and Wright, S. (eds) (1990) *Rural Development: Problems and Practices*, Aldershot: Avebury.

Buttel, F. (1985) Review of Bradley and Lowe's *Locality and Rurality*, *Sociologia Ruralis*, vol. 25, 189–90.

Buttel, F., Larson, O. and Gillespie, G. (1990) *The Sociology of Agriculture*, Westport, CT: Greenwood Press.

Castree, N. and Gregory, D. (2006) *David Harvey: A Critical Reader*, Oxford: Blackwell.

Cloke, P. (1985) 'Whither rural studies?', *Journal of Rural Studies*, vol. 1, 1–9.

Cloke, P. (1997) 'Country backwater or virtual village? Rural studies and the cultural turn', *Journal of Rural Studies*, vol. 13, 367–75.

Cloke, P. and Little, J. (eds) (1997) *Contested Countryside Cultures*, London: Routledge.

Cloke, P. and Moseley, M. (1990) 'Rural geography in Britain', in P. Lowe and M. Bodiguel (eds) *Rural Studies in Britain and France*, London: Belhaven, pp. 117–35.

Cloke, P., Philo, C. and Sadler, D. (1991) *Approaching Human Geography*, London: Paul Chapman Publishing.

Clout, H. (1972) *Rural Geography: An Introductory Survey*, Oxford: Pergamon.

Clout, H. (2002) 'John Terence Coppock', *Proceedings of the British Academy*, vol. 115, 207–24.

Clout, H. (2003) *Geography at University College London: A Brief History*, London: UCL.

Clout, H. (2005) Review of M. Woods, *Rural Geography: Processes, Responses and Experiences in Rural Restructuring, Journal of Rural Studies*, vol. 21, 375–7.

Clout, H. (forthcoming) 'The Auvergne countryside: a retrospective view', in J.-P. Diry (ed.) *Les Étrangers dans les campagnes*, Clermont-Ferrand: CERAMAC.

Cohen, A. (1990) 'The British anthropological tradition, otherness and rural studies', in P. Lowe and M. Bodiguel (eds) *Rural Studies in Britain and France*, London: Belhaven, pp. 203–21.

Colman, D. (1990) 'The development, organisation and orientation of agricultural economics in the United Kingdom', in P. Lowe and M. Bodiguel (eds) *Rural Studies in Britain and France*, London: Belhaven, pp. 165–75.

Cox, G., Lowe, P. and Winter, M. (eds) (1986) *Agriculture, People and Policies*, London: Allen & Unwin.

Crow, G., Marsden, T. and Winter, M. (1990) 'Recent British rural sociology', in P. Lowe and M. Bodiguel (eds) *Rural Studies in Britain and France*, London: Belhaven, 248–62.

Gasson, R. (1971) 'Use of sociology in agricultural economies', *Journal of Agricultural Economics*, vol. 2, 28–38.

Gilg, A. (1985) *An Introduction to Rural Geography*, London: Edward Arnold.

Halfacree, K. (1993) 'Locality and social representation: space, discourse and alternative definitions of the rural', *Journal of Rural Studies*, vol. 9, 23–37.

Hamilton, P. (1990) 'Commentary and introduction', in P. Lowe and M. Bodiguel (eds) *Rural Studies in Britain and France*, London: Belhaven, pp. 225–32.

Harvey, D. (1962) 'Aspects of agricultural and rural change in Kent, 1815–1900', unpublished Ph.D. thesis, University of Cambridge.

Harvey, D. (1963) 'Locational change in the Kentish hop industry and the analysis of land-use patterns', *Transactions of the Institute of British Geographers*, vol. 33, 123–40.

Harvey, D. (1966) 'Theoretical concepts and the analysis of agricultural land-use patterns', *Annals of the Association of American Geographers*, vol. 56, 361–74.

Harvey, D. (2006) 'Memories and desires', in S. Aitken and G. Valentine (eds) *Approaches to Human Geography*, London: Sage, pp. 184–90.

Hubbard, P., Kitchin, R., Bartley, B. and Fuller, D. (2002) *Thinking Geographically: Space, Theory and Contemporary Human Geography*, London: Continuum.

Johnston, R.J. (1991) *Geography and Geographers: Anglo-American Human Geography since 1945*, 4th edn, London: Edward Arnold.

Jones, G.E. (1973) *Rural Life*, London: Longman.

Lowe, P., Bradley, T. and Wright, S. (eds) (1987) *Deprivation and Welfare in Rural Areas*, Norwich: GeoBooks.

Marsden, T.K. (1988) 'Exploring political economy approaches in agriculture', *Area*, vol. 20, 315–22.

Marsden, T.K. and Little, J. (eds) (1990) *Political, Social and Economic Perspectives on the International Food System*, Aldershot: Avebury.

Marsden, T.K., Munton, R.J.C., Whatmore, S. and Little, J. (1986) 'Towards a political economy of capitalist agriculture: a British perspective', *International Journal of Urban and Regional Research*, vol. 10, 498–521.

Marsden, T.K., Whatmore, S.J. and Munton, R.J.C. (1987) 'Uneven development and the restructuring process in British agriculture', *Journal of Rural Studies*, vol. 3, 297–308.

Massey, D. (1985) 'New directions in space', in Gregory, D. and J. Urry (eds) *Social Relations and Spatial Structures*, London: Macmillan, pp. 9–19.

Midmore, P. (1996) 'Towards a postmodern agricultural economics', *Journal of Agricultural Economics*, vol. 47, 1–17.

Milbourne, P. (ed.) (1997) *Revealing Rural 'Others': Representation, Power and Identity in the British Countryside*, London: Pinter.

Munton, R.J.C. and Goudie, A. (1984) 'Geography in the United Kingdom, 1980–1984', *Geographical Journal*, vol. 150, 27–47.

Murdoch, J. and Pratt, A.C. (1993) 'Rural studies: modernism, post-modernism and the post-rural', *Journal of Rural Studies*, vol. 9, 411–27.

Murdoch, J. and Pratt, A.C. (1994) 'Rural studies of power and the power of rural studies: a reply to Philo', *Journal of Rural Studies*, vol. 10, 83–7.

Newby, H. (1977) *The Deferential Worker: A Study of Farm Workers in East Anglia*, London: Allen Lane.

Newby, H. (1980) *Green and Pleasant Land? Social Change in Rural England*, Harmondsworth: Penguin.

Newby, H. (1983) 'The sociology of agriculture: toward a new rural sociology', *Annual Review of Sociology*, vol. 9, 67–81.

Newby, H. and Buttel, F.H. (1980) 'Toward a critical rural sociology', in F.H. Buttel and H. Newby (eds) *The Rural Sociology of the Advanced Societies: Critical Perspectives*, Montclair, NJ: Allanheld, Osmun, pp. 1–35.

Newby, H., Bell, C., Rose, D. and Saunders, P. (1978) *Property, Paternalism and Power: Class and Control in Rural England*, London: Hutchinson.

Pacione, M. (1984) *Rural Geography*, London: Harper & Row.

Pahl, R.E. (1965a) 'Trends in social geography', in R.J. Chorley and Haggett, P. (eds) *Frontiers in Geographical Teaching*, London: Methuen, pp. 81–100.

Pahl, R.E. (1965b) 'Urbs in rure', Geographical Paper No. 2, London: London School of Economics.

Pahl, R.E. (1966) 'The rural–urban continuum', *Sociologia Ruralis*, vol. 6, pp. 299–329.

Phillips, D. and Williams, A. (1984) *Rural Britain: A Social Geography*, Oxford: Blackwell.

Philo, C. (1992) 'Neglected rural geographies: a review', *Journal of Rural Studies*, vol. 8, 193–207.

Philo, C. (1993) 'Postmodern rural geography? A reply to Murdoch and Pratt', *Journal of Rural Studies*, vol. 9, 429–36.

Stoddart, D.R. (1986) *On Geography: and Its History*, Oxford: Blackwell.

Thomson, K. (2001) 'Agricultural economics and rural development: marriage or divorce?', *Journal of Agricultural Economics*, vol. 52, 1–10.

Urry, J. (1984) 'Capitalist restructuring, recomposition and the regions', in T. Bradley and P. Lowe (eds) *Locality and Rurality*, Norwich: GeoBooks, pp. 45–64.

Whatmore, S. (1991) 'Agricultural geography', *Progress in Human Geography*, vol. 15, 303–10.

Whatmore, S.J., Munton, R.J.C., Marsden, T.K. and Little, J.K. (1987) 'Towards a typology of farm businesses in contemporary British agriculture', *Sociologia Ruralis*, vol. 27, 21–37.

Wibberley, G. (1960) 'Changes in the structure and functions of the rural community', *Sociologia Ruralis*, vol. 1, 118–27.

Woods, M. (2005) *Rural Geography: Processes, Responses and Experiences in Rural Restructuring*, London: Sage.

2 Agency and environment in the transition to a productivist farming regime in England and Wales, 1939–50

Brian Short

Introduction

Ecocide, or environmental destruction, and the attack on culturally significant buildings, have recently been highlighted as purposeful elements in wartime tactics (Bevan 2006). But another form of wartime environmental impact has been rather less studied: the transformation of landscapes to meet the requirements of defence or for the supply of emergency food supplies. By late 1939, British preparations for a siege economy propelled the countryside into greater productivity for human food, aiming to divert shipping away from food imports to the importing of the machinery and munitions of war. Such emergency strategies took immediate precedence over the fledgling rural amenity, planning and aesthetics movements.

The agricultural sector's response to wartime requirements drove farming towards greater commercial and technical modernity. Land use was one measure of the change. Thus, in 1939, arable in Suffolk as a percentage of grass and arable was 69.7, a high figure even then, but by 1945 this had risen to 80.3 per cent. In the same period Kent's arable area increased by 174,000 acres, overtaking grassland in 1941 for the first time since 1892 (Trist 1971: 179; Garrad 1954: 4).[1] Although the concepts of productivism and post-productivism are currently disputed as macro-level descriptors of agricultural change, there can be little doubt that the period 1939–50 witnessed many policy, attitudinal and landscape changes which underpinned the following agricultural productivist boom for a generation (Wilson 2001; Evans *et al.* 2002; Walford 2003).

By 1939 the countryside resembled an undercultivated wilderness, producing just one-third of the United Kingdom's food requirements. The impact of interventionist ad hoc government policies since the early 1930s varied with locality, but the area under rough and hill grazings had grown to reach unprecedented levels as arable was abandoned and poorer grasslands deteriorated. Thin downland soils were often little more than rabbit warrens, while understocked clays fell to tussock grass, thorns, briar and scrub. However, a key component of change was initiated in September 1939 with the activation of the County War Agricultural Executive Committees (CWAECs) by the Ministry of Agriculture, charged with ensuring wartime sufficiency of food. In such an intervention 'the State virtually became a partner in Britain's oldest industry' (Ashby and Evans 1944: 139). Many actually saw the war as heralding a return to better times. Lord Cornwallis, chair of the Kent WAEC, wrote in 1944:

> Anyone who knew his Kent five years ago will now see a very different picture. Once more does the land begin to look as if it was cared for and once more really worthy of being regarded as 'The Garden of England'.
>
> (Cox 1944: 123)

And the editor of the *Farmers Weekly* wrote in December 1939 that it was now 'back to good farming ... to the methods our grandfathers knew and to the crops they grew [but now with] all the advances of science at our disposal'.[2]

This chapter concentrates on these county committees, their powers and their interaction with the rural societies and environments of England and Wales. It seeks to place their role and this period within the context of the origins of productivism, and thereby add a necessary historical-geographical dimension to further debates on productivism, demonstrating in particular that the extent of these early state-directed uniformities of policy can be contrasted with a spatial fragmentation of outcomes, as the particularities of rural places were recognized and indeed enhanced through the new structures of surveillance and control.

Control of farming

Following the success of wartime county committees in increasing food production during the First World War, the resurrection of a similar structure was mooted during the 1930s. Indeed, by 1936, chairmen, executive officers and secretaries for each proposed county committee had been selected by Lord Lieutenants, who frequently turned to members already serving on county council agricultural committees. Leading members of the Kent WAEC met unofficially at a 'memorable small private party' in a Maidstone hotel late into the night at the time of the Munich crisis in September 1938 (Cornwallis 1942: 76). In May 1939 the Agriculture Development Act aimed to stimulate improvements to grassland and an increase in arable by offering £2 per acre for the ploughing-up of old grassland. The first meeting of the Hampshire WAEC was held on 30 June in Winchester, with a second at the end of August. Formally, the CWAECs assumed local control immediately on the outbreak of war under section 49 of the Defence Regulations (DR). There were 62 CWAECs, each with between eight and 12 members. Separate committees operated for the three Yorkshire Ridings, the three parts of Lincolnshire, the Isle of Wight and the Isles of Scilly. Their establishment was lauded as one of the war's major administrative successes, and possibly the best example of decentralization and democratic use of control (Murray 1955: 338–9). Through sustained efforts, food did remain available, albeit with considerable consumer patience, austerity and hard work by the Ministry of Food in rationing (from January 1940) and distributing supplies.

The CWAECs were strongly placed to influence the precise balance to be struck between tradition and progress. The Minister appointed the members, but there would have been a fairly limited pool of people with the time, experience and ability suitable for membership of the committees, although their unelected nature was later to be the cause of considerable adverse criticism, especially from the National Farmers Union. The chairmen, according to a written answer in the House of Commons in January 1941, had to possess a knowledge of local agricultural conditions, the 'confidence of the local agricultural community', administrative ability and 'a strong sense of public duty'.[3] Despite the waning power of rural elites and the democratization of rural politics in the twentieth century (Woods 2005: 23–46), the CWAECs comprised influential landowners, farmers and land-related personnel, appointed to impart urgency and good agricultural practice, and all imbricated in discursive power relations and webs of influence, control and surveillance. The chair of the Bedfordshire WAEC, for example, was H.J. Humphreys JP, who had a large farm on the Duke of Bedford's estate near Woburn. The Northants WAEC was chaired by H.R. Overman from Brampton Ash, Market Harborough, from a long line of respected Holkham estate Norfolk farmers. By 1952 he was chair of the Farmers' Club of London. The Huntingdonshire WAEC was chaired by Major R.G. Proby until 1943, and his committee included the Ninth Earl of Sandwich from Hinchingbrooke House, former Assistant Private Secretary to Stanley Baldwin and Lord Lieutenant in 1926. A photograph *c.*1926 shows all three men together at Hinchingbrooke.[4] In the Soke of Peterborough the chair until June 1942 was William Cecil, Fifth Marquess of Exeter, from Burghley, Stamford. One 'innate shire Tory' but also a fascist agriculturalist was the Earl of Portsmouth, a vice-chairman on the Hampshire WAEC. Despite his promotion of extreme right-wing ideas on traditional husbandry and self-sufficiency, he was not interned, but served on the committee throughout the war (Portsmouth 1965: 77–96). The list of such elite country personnel could go on: perhaps we might leave it with Captain Charles Fitzroy, the Duke of Grafton from Euston Park, vice-chair of the West Suffolk WAEC. Highly active, he oversaw from 1942 the conversion of poor acidic parkland grazing and clearance of 100 acres of bracken and thorn for corn and lucerne for sheep feed. He also leased 550 acres to the West Suffolk WAEC, who cleared and ploughed for barley, wheat and oats, all this helped by 130 Dr Barnardo's boys billeted in the house, who worked away at pulling bracken (Trist 1971: 146–7; Grafton 1943: 85–9, 1947: 127–9).

The Welsh situation was slightly different. Landed elites were less prominent, although the Brecon MP, W. Jackson, still complained that the committee members were too elderly and were selected because of their wealth and status rather than because they were 'the best in the industry'. Whatever the truth of this assertion, perhaps typical was Alderman Thomas Thomas, JP, long-time chairman of the Llanelly and District Farmers Society since its inception in 1908, and a prominent figure in the agricultural community. He was a member of the Carmarthenshire WAEC, chair of the Llanelly District Committee as well as four other subcommittees, and a member of several others. The minutes of the Glamorgan WAEC for 12 September 1939 noted that subcommittee chairmen were to be appointed from within the executive committee, and such chairmen would then approach those 'who in his judgement, were willing and capable to serve with him'.[5] Clearly, then, overlapping local policy networks could form as friends, acquaintances and kin were engaged.

The powers of the CWAECs grew monthly. Sweeping DR legislation allowed the requisition of part or whole properties; enforced surveillance through farm visits and the administering of the National Farm Survey (1941–3), which among other things graded farmers on their abilities as 'A', 'B' or 'C'. CWAECs could enter and inspect land, and control virtually every aspect of farming (Short *et al.* 2000). However, many prided themselves on their ability to get the best out of their farmers without recourse to their full powers. But everything was improvised and rushed, and mistakes were made in the drive to maximize productivity, and especially in the early war years in the conversion of grassland to arable. This massive and swift structural realignment necessarily affected other sectors adversely, especially poultry and pigs, and to a lesser extent cattle and sheep. Only the dairying sector retained its levels of output, thanks to a somewhat belated recognition of its merits in wartime. The Minister for Agriculture, R.S. Hudson, appointed by Churchill in May 1940, toured the country, urging on the plough-up. By June 1940 he was pushing the CWAECs for an extra 1.5 million acres of arable (Hurd 1951: 20).

The multitude of CWAEC tasks was carried out through devolved subcommittees. District committees were at the forefront in linking Whitehall to individual farmers: primarily organized around rural districts, they comprised from four to seven volunteers in good standing and with a good knowledge of local farming. Over England and Wales, 453 of these committees operated in the early war years, increasing to 478 in the later years. A large county such as Devon had 17 such committees, by 1943–4 involving 211 members, mostly farmers. There was also an extensive network of specialist subcommittees undertaking such functions as the allocation of plough-up orders to meet county quotas, set at approximately half the area lost to the plough since 1918, and the payment of the ploughing subsidy. The range and quantity of work increased as the war progressed, with what John Moore called 'a fat beribboned bundle of documents' frequently changing hands from one committee to another (1948: 159–63). Tasks also included the distribution of tractors and other equipment; liaison with the armed services; direction of temporary sources of labour, such as the Women's Land Army, the Timber Corps and prisoners of war; the encouragement of land drainage; pest and disease control; and the provision of accommodation. The precise numbers of such committees varied from county to county, and from one year to another. Altogether there were 771 recorded in the Ministry of Agriculture's archive.[6] However, this is almost certainly an underestimate, since not all the war years' activities are preserved, and committees were fluid, with some being of fleeting appearance and others changing their names.

The functions of such subcommittees were initially assigned by Ministry officials, but they also came to reflect the particularity of individual counties. Cheshire's important dairying and Cornwall's horticulture both necessitated complex arrangements. In Middlesex there was an emphasis on market gardening; Northumberland had to administer farms within the Redesdale artillery range, necessitating the appointment of three army officers; Lancashire had a specialist marginal land committee; Cumberland a commons improvement committee for a short period; and Durham a stints committee to control the amount of grazing on out-bye common land. Hampshire had a New Forest pastoral development committee, and Wiltshire even had a meat pie subcommittee, reflecting the important pork pie and bacon factories of Bowyers and Harris.

The spatial expansion of early productivism

In response to CWAEC and farmers' efforts, the harvest of 1943 was the most spectacular for decades, but by this time corn crops had been taken for several years in succession on many lowland fields, and it was clear that yields would henceforth fall. Light lands were becoming exhausted and heavy lands weed-ridden. Five million acres of pasture had already been ploughed in Britain as a whole, but now new land was required. If tillage were to be further expanded, it became clear to most that marginal hill country, furze-covered sands and gravels, fens and waterlogged clays must be further exploited. In 1941 the Select Committee on National Expenditure felt that 'concentration on ploughing fresh land led to the neglect of measures for much needed improvement of existing pasture and arable' (Select Committee 1941: 24). And in the uplands, problems had materialized early in the war as lowland farmers converted pastures to arable, thereby necessarily reducing purchases of hill stock. Now drainage, lime, phosphate, new roads and fences would be required, as well as supplies of labour, though it was recognized that the low yields and high costs of such operations would be entirely uneconomic (Whetham 1952: 95–6). But the expansionist mood was relentless. As early as December 1940 a subsidy for each ewe carried on a hill sheep farm had been introduced, becoming an important funding consideration on these holdings, and joined by a hill cattle subsidy in 1943 and marginal production grants. Areas such as Swaledale, otherwise largely unaltered in their farming systems, benefited (Long and Davies 1948: 43–7), and much was later encompassed in the 1946 Hill Farming Act. Elsewhere, 50 per cent grants for draining farmland were offered for ditching, mole draining and tile draining, and in 1941 alone, 40,000 schemes were approved in England and Wales, covering well over a million acres, with nearly two million acres tackled in the next two years.

These less productive environments were joined by others, many of which were affected by the 'Dig for Victory' campaign, which urged that every possible space be used. Photographs, free newsreel films and newspapers related how urban, suburban and peri-urban spaces were put to the production of food: Bristol had more than 15,000 allotments by the end of May 1941, and food now came from public parks, squares, recreation grounds, football pitches, railway banks, promenades and flower beds, the British Museum forecourt, and even from bombed ruins, after the ubiquitous rosebay willowherb had been cleared. The government sent out exhortatory notes to be used by clergy in their sermons. Allotment and gardening societies, pigfeed clubs, school activity clubs and all manner of other informal groupings sprang up. The MP Rupert De La Bère (Evesham) was especially vehement in his call for golf course fairways to be ploughed, citing an estimated 550 miles of 'unproductive fairways' as a scandal.

> Is it not easier to cultivate land which has already been drained than to plough up the slopes of Plynlimon and to endeavour to cultivate derelict and waterlogged land? Why not plough up the drained fairways of the golf courses?

Hudson replied simply that such a policy would not be in the national interest. Fox hunting also continued on the basis that it was exterminating foxes, although some destruction of crops by horses and hounds was reported.[7]

With profound optimism the CWAECs addressed some of the poorest land – neglected, under-farmed, covered with bushes, waterlogged or waterless, lacking roads and buildings. Thus, production extended on to commons, detached pieces, footpaths, derelict building sites or where grassland farmers lacked experience or knowledge and handed over land voluntarily, as on Romney Marsh to the Kent WAEC. In East Sussex the Forestry Commission yielded up most of its plantation at Friston, and this was merged with adjacent plots to form a 1,826-acre holding, farmed by the WAEC. Of South Staffordshire it was noted that 'the holdings in this district are mainly amongst pit and slag heaps and there are plots of uncultivated land'. Liverpool's urban fringe contained farms already scheduled for urban development and therefore lacking investment. These were invariably graded as 'C' farms by the Lancashire committee (Rawding 2007). At Peacehaven, on the Sussex downland, undeveloped plots with untraceable owners were requisitioned, and there were many owners who eventually located their plots, hoping to build a bungalow, only to find it producing cabbages![8]

Agricultural expansion frequently competed with other critically important land uses. There was competition for East Anglia's flat terrain for airfields, the War Office had to intervene in a dispute over the militarization of much of the South Downs, and military use of moorland and heathland now intensified. Indeed, by 1944 a remarkable 20 per cent of the total land area of Great Britain was in military use, and by 1945, although the plough-up campaign had created about six million acres of arable land in Britain, farming had still lost a net half million acres, much to military uses.[9] Many changes were temporary, and although military occupation for airfields and defence lines, and as preparation for D-Day, for example, had left nearly 20,000 landscape features, many commandeered areas reverted to farmland post-war (Foot 1999, 2006, 2007). One-third of the Yorkshire Wolds was taken over in June 1943 as a training ground, with sheep being removed, hedges flattened by tanks, and crop acreages falling by 40 per cent (Long 1969: 35). Today, all traces of this are gone, as are the modified landscapes of Slapton Ley in Devon, where the Americans prepared their invasion of Normandy, but sombre memorials honouring those who died during these preparations remain.

The reclamation of fens and moorlands

To particularize the geographies of change, two broad habitat types will be examined: fen/marshes and moorlands (Figure 2.1). Both have undergone long-

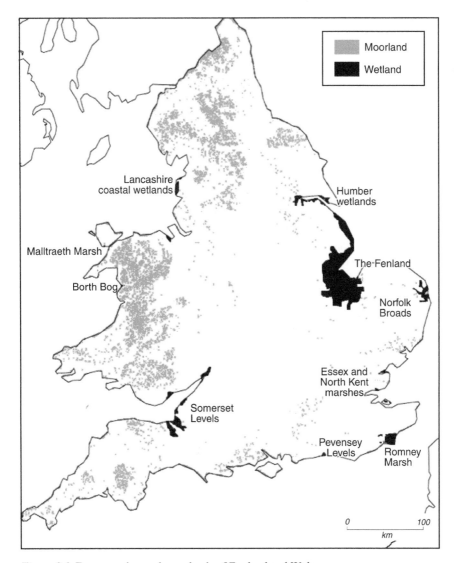

Figure 2.1 Fens, marshes and moorlands of England and Wales.

term cyclical farming changes, with historically greater or lesser amounts of activity within their boundaries, depending on both socio-economic and physical fluctuations. But now the emphasis was on cultivation. Indeed, reading the accounts of reclamation and the plough-up campaign in the pages of the *Journal of the Royal Agricultural Society* or *Agriculture: The Journal of the Ministry of Agriculture* or the *Farmers Weekly* and farming press, one is struck by heroic, triumphant, almost pioneering accounts of overcoming nature. Laurie Lee, despite disliking the prose he himself wrote for the Ministry of Information's *Land at War*, proclaimed:

> From coast to coast now the ploughs were fanning out over new ground. They were climbing the lower Pennines, breaking the Devon moors, sweeping long brown paths over the Berkshire Downs, and threading their way between the derelict bungalows of Sussex holiday camps.
>
> (Ministry of Information 1945: 20)

Fens and marshland

By 1939 the many areas of wetland were often poorly maintained. But the potentially highly productive ecosystems of the East Anglia fenland were a clear target for the committees, particularly those on the eastern edges, such as Feltwell, Swaffham and Burwell Fens, that had escaped previous conversion to farmland. The wetland habitat was now to be sacrificed, although this was by no means a straightforward task (Ennion 1942).

When Alan Bloom, the nurseryman and writer, purchased land in the north of Burwell Fen in 1938 it was largely under water, but his wartime efforts with ditching gangs and draglines turned his farm into a productive unit, with the effort recorded in the 1941 film *Reclamation* and a 1944 book. The King and Queen, R.S. Hudson, the Duke of Norfolk and WAEC members all came in 1942 to inspect the reclamation (Bloom 1944: 160–2). At Swaffham and Burwell, concrete roads – the essential key to improvement – were laid across the fen in 100-acre squares and the land within was then reclaimed – every 100 acres requiring five miles of drains. Ditches were cleaned and strengthened, and the land slowly dried out. The extensive reed beds were then cleared, partly through the efforts of about 60 Women's Land Army members, and then caterpillar tractors with Canadian prairie-busters, disc harrows and Australian stump-jumper ploughs broke up the land to make it ready for cultivation. The old bog oaks, numbering as many as 50–60 per acre, seven feet thick and 100 feet long or more, were dynamited out where necessary (Figure 2.2).

Bloom's account of his reclamation of Burwell Fen, shot through with triumphant allusions to progress, war effort and fighting nature, recalls H.C. Darby's Whiggish accounts of earlier drainage, published incidentally during the same war years (Darby 1940a, b). But the arch-reactionary writer James Wentworth Day had purchased part of Burwell Fen in the 1930s for conservation and duck shooting, and his account was correspondingly very different. Returning

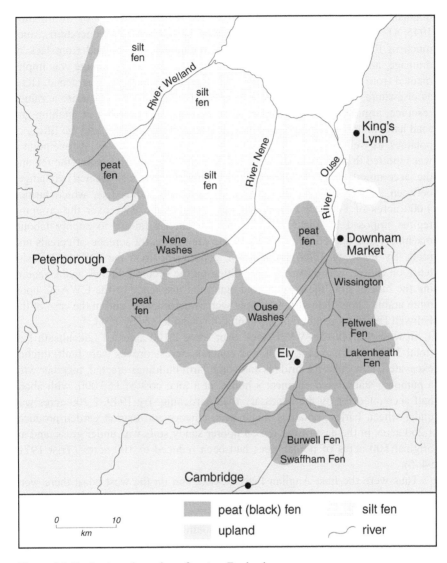

Figure 2.2 Fenlands and marshes of eastern England.

during the war for a visit full of nostalgia for shooting and fishing days, he now found it drained by the CWAEC, whose concrete road was 'a symbol, cold and hard, of the mechanical values which slew that last sanctuary of reed and bird, flower and fish'. Now 'a caterpillar tractor pulled a four-furrow plough. Great slices of chocolate-coloured soil turned relentlessly over behind the spitting, chuffing passage of the tractor' (Wentworth Day 1943: 167–8). By 1942 the Fen was producing potatoes, mangolds and beet, and the WAEC was reputedly farming about 7,500 acres in the area.[10]

The Norfolk WAEC similarly reclaimed 1,500 acres of Feltwell Fen for triumphant harvests of wheat, potatoes and sugar beet (Ministry of Information 1945: 41–4). A specialist subcommittee was formed to oversee operations, since much of this potentially fertile land had remained undeveloped from lack of drainage, and desolate because of inaccessibility. But now drainage was implemented from March 1941, prisoners of war deployed, and roads hardened. Here, as elsewhere, the prisoners of war, together with Irish workers, were a valued resource, argued over and switched from job to job. Feltwell Fen ditching and road hardening had to give way to the demands of the military and the lifting of potatoes, as well as sugar beet for the nearby Wissington factory. At one point it was reported that 30 Irishmen were sent to Feltwell, but, after seeing the fen and the accommodation, left the same day! There were also heated exchanges between the Norfolk WAEC and the Weasenham Farms Co., which farmed 4,000 acres of Feltwell Fen throughout much of the war, over the cropping regime imposed by the CWAEC, and the latter's failure to supply labour, machinery and barbed wire to harvest the imposed extra acreage of cereals and roots. By February 1944, soil erosion, consequent upon the CWAEC's insistence on tillage, was occurring and silting up the scoured ditches again, requiring the hire of draglines. Again and again the insistence of the CWAEC upon more arable rather than grassland was seen as detrimental, and in the case of the Feltwell Fen Farm resulted in substantial financial losses by 1944.[11]

In west Suffolk the WAEC took over some 2,100 acres of Lakenheath Fen (total 3,700 acres). Again, chalk and concrete access droves were built, ditches excavated, new culverts, bridges and some farm buildings erected, together with a pumping station and engineer's house, at a total cost of £60,000, with about half accounted for by the necessary road hardening. By 1949, 2,200 acres was under wheat, barley, potatoes, sugar beet, chicory and market garden produce; 1,250 acres in the east of the fen on poorer sandy soils was under grass, and an original 500 acres of poplar trees had been reduced to 100 acres (Trist 1971: 94–5).

Thus were the East Anglian Fens tackled. But on the west coast there were comparable large-scale reclamation projects, of which three can be noted here: Frodsham Marshes (Cheshire), Malltraeth Marsh (Anglesey) and Borth Bog (Cardiganshire). Frodsham Marsh, on the southern banks of the Mersey estuary, covered 3,100 acres and, fragmented between 86 different owners, was used for the summer grazing of dry and young stock. It was one of three such marshes tackled by the Cheshire WAEC, the others being Burton Marsh, where barley, clover and early potatoes were successful, and Whitley Reed, where potatoes and swedes were eventually successful despite the prevalence of bog oaks within the moss (Mercer 1963: 180). Frodsham was ill-drained and subject to flooding, with ditches choked with weeds because of the costs of clearance during the depression and with occupation roads poorly maintained. But the decision was taken in 1942 to tackle 1,500 acres of the marsh. It was therefore requisitioned, new ditches cut, pumps installed, roads built, tall hedges cut back or grubbed out and 200 miles of fencing (mostly barbed wire) inserted. Dragline excavators deepened many miles of subsidiary ditches. All was done with great speed and in the face of resource shortages. Italian and German prisoners of war and Irish workers were again employed, and over 4,750 tons of lime applied. By 1945 there were 944 acres of tillage and 772 acres of reseeded grass, and after the war it went to the Agricultural Land Commission rather than being returned to fragmented ownership (Carr and Mercer 1947). Both Frodsham and Burton Marshes appear to have repaid the huge wartime investments – of the latter, an executive officer for Cheshire WAEC later wrote: 'Land reclamation during the war was usually a dismal business judged from the standpoint of costs and returns. In this respect Burton Bog will provide historians with an exception to prove the rule' (Mercer 1963: 149–50).

Malltraeth Marsh certainly illustrated the 'dismal business'. These 4,000 acres of reclaimed alluvium at the mouth of the River Cefni, Anglesey, had been improved by late eighteenth- and early nineteenth-century embanking and flood protection schemes. Wartime improvements were now urged by the Ministry, to be overseen by the Anglesey WAEC, amid concerns from local farmers that the financial outlay would not be recouped in any enhanced grazing value. Reclamation in 1914–18 had not been maintained and the marsh had reverted to a primeval state. Bitter disputes arose between the drainage engineer, the CWAEC, particularly its executive officer, Griffith Jones, and what was described as a 'poverty-stricken and unenterprising' catchment board over the efficiency of the drainage of the marsh and its cultivation. By March 1942 the CWAEC itself was riven by arguments, culminating in the chairman's resignation from a Malltraeth Marsh subcommittee. Despite delays in draining this difficult basin, and consequently in producing cereals, the Ministry urged the seemingly dilatory WAEC on, with the result that a rift also became apparent between Ministry officials and the WAEC members, who were now denying responsibility for the scheme altogether. The exceptionally wet harvest period of 1943 was illustrative. Most of the cereals were lost after flooding forced the abandonment of cutting, although several tenants seem to have made little attempt to cut and stook the weed-ridden corn, and some complained of a lack of timely machinery help from the WAEC. Lord Anglesey, highly influential in the area, requested a meeting with Robert Hudson and local MP Megan Lloyd George to discuss the issue. But by April 1944 there were 80 owner-occupiers of the marshland, who

> appear to take very little interest in the drainage scheme as the ditches which were reopened and thoroughly cleaned two years ago are now fouled with weeds and silt. If some action is not taken immediately to maintain this work the whole marsh will again become derelict.

With labour supplies also difficult, by August 1944 one observer reported 'a shabby expanse of rushes, weeds and scrub, with the ditches marked mainly by the fact that the weeds in them are taller and more flourishing'. About half the marsh was seen as derelict, and 'the failure of the Anglesey committee to see that full agricultural use is made of the drainage improvements which have been secured is nothing short of a disgrace'. Some committee members, such as Lord Anglesey, were interested parties in the marsh (to the fury of the chairman); some had always been antagonistic, suffering from what was referred to as a 'marsh complex' in feeling that the marsh would never be of use. Approximately £100,000 was spent on the project during the war, more than double the original estimate, and the WAEC was still working on this by late 1945, using Italian prisoners of war and Women's Land Army labour, and still convinced that the area could be rendered more productive, given a good drainage and water supply. By the end of the scheme, in 1947, £147,000 had been spent.[12]

A final wetland example comes from the Borth Bog (Cors Fochno), south of the River Dyfi, Cardiganshire, also with its own WAEC subcommittee. This large raised mire, drained since the 1820s for grazings, again had a complex pattern of landownership and also common rights to cut peat. Based on earlier reports of the potential for reclamation, large-scale drainage schemes were implemented by the WAEC. Commentators included C.S. Orwin on behalf of the Land Reclamation Society, who also warned that any such scheme would necessitate land being 'taken by compulsion' because of the many small owners who gained income from peat diggings. This action again ran counter to local feeling that the high maintenance costs, in relation to productivity, would cause its abandonment once peace came.[13] A 1941 report considered that local farmers had 'neither initiative nor the necessary knowledge to keep the works that do remain in safe condition.... They wait until complete failure [of the river embankments] takes place and then either do nothing or put in a patch that has no chance of lasting' (Sheail and Mountford 1984: 45). Nevertheless, attempts to cultivate parts of the northern edge of the bog were undertaken, but some did indeed lapse post-war, although by the 1960s a considerable area of farmland was still operational.

Moorlands

The inhospitable moorlands presented another challenge to expansion. Their tops were frequently peat-covered, and the potentially cultivable land was enclosed on the lower slopes, where the edge of cultivation had fluctuated over time with climatic and economic change. On the good in-bye land of the Cheviot Hills, intensification entailed reducing the length of leys, and elsewhere on the moorland edge ploughing and reseeding allowed denser sheep stocking (Pawson 1961: 4–9 and *passim*). But on the higher slopes the CWAECs had to make judicious choices. Would the returns from reclamation be worth the huge level of inputs, however much was at their disposal? And could they persuade farmers to undertake cultivation of slopes that defied normal principles of good husbandry? Certainly hundreds of upland farmers felt that unattainable ploughing quotas were being imposed, a view shared by the NFU, landowners and even CWAECs themselves. Both Cardiganshire and Carmarthenshire WAECs were concerned that soil impoverishment would surely and swiftly follow (Moore-Colyer 2005: 571–2).

The first large-scale reclamations of the war were initiated by the Montgomeryshire WAEC at Long Mountain and by the Ministry of Agriculture itself at Dolfor in spring 1941. Work started with 85 acres on Long Mountain, where bracken and thorns were uprooted using steel hawsers. The land was then readied for seed potatoes but the crop failed, with bracken roots a problem, and a pioneer crop should have been sown first, a point stressed by Stapledon (1943: 103), so that cattle could manure and trample the ground. The proponents had been too impatient. In 1941 they ploughed deep again after getting rid of bracken, and now reclaimed another 400 acres of Long Mountain at higher altitude, spread lime and slag, and this time sowed rape for sheep, in time for potatoes by spring 1942. At Dolfor the first 1,000 acres was ploughed in May 1941 using five huge lease-lend Allis-Chalmers HD.7 caterpillar tractors and eight other track-laying tractors to haul ploughs, leaving behind 1.5-mile furrows, which subsequently yielded 'the biggest potato patch in British history' (Ellison 1943: 100–11; Ministry of Information 1945: 45–6; Harvey 1995; West 1941). Thereafter a further 1,000 acres a year was tackled in 500-acre blocks, using crawler tractors and Massey-Harris prairie-buster ploughs. The tufted herbage could not be burned off because of blackout restrictions but by the end of 1942 about 6,000 acres of hill land between 700 feet and 1,700 feet initially covered by bracken, nardus and nardus-fescue, and molinia had been tackled. The productivity of the WAEC uplands was reflected in increased stocking densities of cattle and sheep by 1944 (Ellison 1946: 54–67). However, the executive officer's efforts to secure petrol coupons and transport for the delivery of the lime (five tons per acre being required) generated more than 800 letters, forms and other documents (Moore-Colyer 2005: 566). The impact on the land use and stocking of Montgomeryshire, a county with a large moorland area, and its upland parishes of Llanfihangel and Kerry (a large parish containing Long Mountain and Dolfor) is summarized in Table 2.1.

These moorland reclamations could involve tackling otherwise uneconomic steep slopes: on part of Holcombe Moors, near Bolton, gradients of 1:3 were being ploughed and reseeded by February 1944 (Ward 1988: 55). Approximately 10,000 acres of Pennine moorland was ploughed for kale, oats and cattle fodder (Ministry of Information 1945: 47), and in Cheshire 400 acres of semi-derelict, acidic molinia on Wildboarclough was reclaimed by the WAEC using large amounts of lime, with walls being repaired, and 200 acres of leys established. At great cost, the stock-carrying capacity was doubled (Mercer 1963: 163–4). In Devon, by 1943 it was admitted that mistakes had been made through ignorance of the special needs of moorland areas, or through excessive zeal, but by the end of 1942 about 11 per cent of the county's total rough grazings had been reclaimed, thanks primarily to huge inputs of labour, either by the WAEC or by individual farmers (Horne 1943: 93). The county of Northumberland comprised approximately one million acres, of which more than 40 per cent was rough grazings. Farming here was handicapped by a shortage of lime and by poor drainage. By 1939

Table 2.1 Agricultural change 1939–45 in Montgomeryshire and in two upland parishes (acreage figures rounded to nearest whole number)

	Montgomeryshire		Llanfihangel-yng-Ngwyfa		Kerry	
	1939	*1945*	*1939*	*1945*	*1939*	*1945*
Wheat acreage	2,991	9,110	12	73	313	745
Barley acreage	1,750	6,162	28	28	50	445
Oats acreage	15,839	25,234	477	835	1,083	1,501
Potatoes acreage	888	3,821	27	102	37	212
Clover, sainfoin, etc. acreage	24,932	46,023	885	1,923	1,616	4,184
Permanent grass acreage	204,315	140,166	5,935	4,142	9,126	6,545
Regular male workers	2,321	2,649	49	60	142	152
Women and girls	509	499	13	12	30	18
WLA	–	127	–	1	–	5
POWs	–	399	–	6	–	31
Casual workers	565	658	13	14	22	30
Total workers	3,871	4,464	93	95	194	243
Dairy cows and heifers	29,793	35,076	569	985	1,366	1,550
Total cattle and calves	83,008	94,358	2,467	2,968	4,690	5,790
Total sheep	595,419	587,289	13,310	12,674	25,401	25,862
Total horses	11,263	10,066	327	281	728	572
Total pigs	23,869	18,025	559	544	879	782
Total poultry	539,047	546,355	8,793	10,639	27,153	29,911

Source: TNA: MAF 68/3931, 4153

Note
Acreage figures are rounded to the nearest whole number.

there was only one combine harvester in the county, and one farm dryer, but by 1944, numbers had risen to 49 and 18 respectively. With the application of lime and compound fertilizers, and the work of over 70 contractors with more mechanization, especially grain dryers, the overall increase in tillage was threefold (Pawson 1961: 4–9).

The impact on rural society

The enforced agricultural modernity implicit in the CWAECs' work cannot be divorced from its social context and impact on individual agency, given the significance of the war in many people's self-definition. Many of the official accounts reveal little about agency, being 'a sort of denatured history [in which] exciting clashes of personality are most discreetly veiled ... in ... careful analysis' (Bury 1954: 509). We therefore turn to a brief discussion of the impact of the work on the people most affected. In general, the early months of the war saw rural communities unchanged, except for those receiving middle-class and child evacuees, but summer 1940 brought the fall of France, rumours of invasion, the removal of many evacuees again to Cornwall and Wales, and air raids in the eastern and south-eastern countrysides. The evacuees provoked some overcrowding and even hostility in some villages, undermining the myth of a united 'people's war' (Howkins 1998: 82–3; Rose 2003; Sheridan 1990: 55–68).

Many farmworkers had already found jobs elsewhere during the Depression, and as work on military construction sites increased by 1939, many feared that labour supplies would be insufficient to achieve any planned arable expansion. As was said of Devon at the end of the war: 'To lose 10,000 men and then be asked to double the arable acreage of the county is rather a tall order' (Hayter Hames 1947: 88). All agricultural workers were therefore required, under the Emergency Powers (Defence) Regulation of May 1940, to remain on the land – for a minimum wage of 48 shillings per week. Farmers protested at these wage increases. Even so, seasonal labour requirements remained especially pressing, and a great variety of new arrivals now appeared in the rural community (Moore-Colyer 2006): emergency land corps or voluntary land clubs, Women's Land Army members (67,000 by 1944), volunteer army and air force helpers, conscientious objectors, men drafted from labour exchanges, and, from 1941, Italian and German prisoners of war (reaching 57,000 by 1945). Accommodation for all these extra workers also presented its own problems, and all manner of hostels were pressed into service. On the Dolfor scheme, 17 tractor drivers, newly trained, were engaged, working in shifts from 7.30 in the morning to 10.30 at night, seven days a week, in order to keep the machinery going; a repair gang was on site; and local women were employed on domestic duties in three requisitioned bungalows and three bell tents. The untrained planting gang of about 140 local men, women and schoolboys included refugees.

The impact on rural communities clearly varied with location, farming type and the activities of the CWAECs. Class was also a factor. For the wealthiest families of all, there was the distinct possibility of losing their mansions to the military, since country houses were ideal for military HQs and were rarely treated sympathetically by working-class soldiers, and were sometimes even looted by local people (Robinson 1989).[14] For other wealthier families, life might go on much as before, with the possibility of supplementing rationing with game or paying for black market goods. Farmers also had access to their own produce, and it was the poorer villagers and country town residents who bore the austerity of rationing.

Reactions to the CWAECs were certainly mixed, ranging from high praise to execration. A traditional paternalism associated with many communities might engender compliance with the committee decisions, at least overtly, but elsewhere their actions were met with suspicion or downright hostility. Old friendships might be broken, community reciprocity threatened as neighbours formally judged neighbours; even family ties strained. By far the most controversial action was in dispossessing farmers, mostly 'C' graded, who failed to comply with CWAEC instructions. Over 2,500 farmers were affected either by complete eviction from land and house, having their tenancies terminated, or by having

Table 2.2 The progress of wartime dispossessions

Date	Farm tenancies terminated under DR 62	Cases of taking possession of land under DR 51[a]
July 1942	1,881	4,373
March 1943	2,502	5,631
September 1943	2,661	6,332
May 1944	2,838	6,674
December 1944	2,897	6,739

Source: PD (Commons) replies to parliamentary questions, various years.

Notes

a This figure includes accommodation land, undeveloped and derelict land, common land, sports and recreation land.

Defence Regulation 51 empowered CWAECs to requisition land to promote or increase food production.

Defence Regulation 62 empowered CWAECs to terminate tenancies of land not cultivated in accordance with the rules of good husbandry, extended in 1941 to allow ejection from the holding.

land taken over (Table 2.2). Dispossessions were the subject of considerable discussion at the time, but there was little mention of the impact of dispossession on the farmer within the community, where such an outcome might be seen as a disgrace. As Sir John Mellor told the Commons in 1941: 'His character as an agriculturalist is, indeed, wrecked'.[15] The Minister's response was that, if anything, the committees were too lenient, having known the farmers concerned for many years, and any hardship caused was inevitable in wartime. At least one farmer committed suicide as a result; another was killed by police while resisting eviction; some left the land never to return; others had relatives who sought, and still seek, to publicize what they see as wartime injustices (Farmers' Rights Association 1948; Short 2007). The impact fell unequally upon smaller farmers, who were more likely also to be graded 'C'; or on all-grass or dairy farmers faced with producing arable crops; or by those requiring access to common land that had been ploughed.

CWAEC and Ministry attitudes to affected communities can be illustrated from the fens. The Agricultural Improvement Council considered a 1948 report on the wartime fenland stating that

> [t]he best and most progressive farmers will not live in these inaccessible areas and they tend to become populated by men of the squatter type.... It is not an uncommon thing to meet intelligent and keen young people, in their twenties, who are quite illiterate. [One cannot] expect a high standard of farming from illiterates. The peat-covered fens are not desirable places of habitation for they are in many parts difficult of access. Water supply and sanitation is lacking and the climate is damp and cold in winter. A great deal of subsidence takes place owing to the sinking of the peat, and brick-built houses become displaced and damaged. Their appearance is quite as

bad as in the salt-mining districts of Cheshire. There is a tendency to build dwellings of a more temporary and flimsy nature, which can best be described as shacks. In some fens they present an appalling appearance and from the exteriors are hardly fit for human habitation.[16]

In the peat areas (the Black Fen) there were only the hovels of small farmers whose scattered parcels gave a poor living. Reclamation should therefore progress unhindered, affected residents being given short shrift or seen as unfortunate war casualties. Alan Bloom wrote of his shock at looking beyond his newly purchased farm to see the hundreds of acres 'over which wild nature had completely gained the mastery' (1944: 20). It was acknowledged that 'the fenman is particularly independent and individualistic'.[17] But just as medieval and early-modern reclamations left fen-dwellers, who depended upon reed and peat cutting, fishing and fowling, bereft in the face of 'improvements', so too surely did this latest series of rapid reclamations. Although turf- and litter-cutting barely survived the First World War, and H.C. Darby thought the last turves were cut at Burwell, Reach and Swaffham in 1937, there would undoubtedly have been families still dependent upon the fenland resources for a living (Darby 1983: 234). To some extent, this reliance was replaced by the certainties of a minimum wage, as labour was scarce, and the skills of those who knew how to work on the land were valued again. Thus did men and women work at 'changing the face and purpose' of fenland localities which were seen as 'annoying and useless background[s]' (Bloom 1944: 96).

Cycles of change

Post-war arguments between the uses to which such marginal land as moorlands and fens might be put are well known: the cases for agricultural expansion versus conservation or afforestation repeated in variants across the country. But the pressure upon these environments brought by total war has been less explored. Alan Bloom spent many of the war years reclaiming fenland for farming. But in the midst of his struggles, one day a young man remarked in passing that post-war reversion would be the best target. Bloom was thoroughly annoyed:

> [I]t was staggering to find at such a time – in the midst of the nation's greatest peril – that anyone, or for that matter any body of men, should consider the reclamation of valuable food-producing land a waste of energy, a waste of money, and moreover should calmly contemplate relinquishing it once more to its fate when that immediate peril was past.
>
> (1944: 151)

Such attitudes, helped by the advocacy of the Scott Report (1942) arguing for agricultural development and investment within a sacrosanct farming environment, thereafter offered farmers greater licence under the Town and Country

Planning Act 1947 to change environments as required, all within a structure of feeling emanating directly from the wartime expansion. Muted environmentalist arguments might be heard, but the CWAECs, backed by the full force of government and authorities such as C.S. Orwin, had little time to reflect. The latter wrote in 1945 of

> the further handicap, in many places, of the smallness of the fields ... here and there, landlords and farmers have recognized the need for adjusting their field boundaries to the scope of the agricultural tractor. Some hedges have been removed, others have been straightened, and hedgerow timber, so detrimental to good arable farming, has been reduced.
>
> (1945: 28)

Orwin waged ceaseless campaigns against those he called 'the prophets of the new pantheism'. Meanwhile, the gyrotillers were clearing 100 to 120 yards per hour of oak, ash and elder hedges in Suffolk (Trist 1971: 196). And later, Orwin cites the question posed by a land-girl to an expert tractor-driver: 'What do you think about when you are sitting on your tractor all day?' He replied, 'I looks at the bloody earth and I says, blast it' (1945: 88). So, there is seemingly no overriding aesthetic sense, no rural idyll, and thus probably few senses of loss at the wartime reclamation. No pastoral aesthetic was strong enough politically to withstand the need for more home-grown food, and the overriding agricultural opinion was that such supplies could only be obtained by the rapid conversion of as many farmers as possible to a productivist ethos.

Yet the modern consensus remains that the post-war environmental modifications were 'deleterious if not disastrous', because of intensification and chemical inputs. A familiar litany of damage includes that to ancient monuments and older landscapes such as water meadows or ridge and furrow, loss of hedgerows, pollution, reduced biodiversity and water quality – all in 'a wearying dash for growth' (Bowers 1985: 75–6). It is salutary to reflect therefore that after all the wartime toil undergone in the reclamation of marshland, much of it today has indeed reverted: much of Burwell Fen has been returned to its pre-war wetland habitat (Friday 1997); since 1995 the RSPB[18] has re-created possibly the largest area of fen habitat in the United Kingdom at Lakenheath Fen; and today Malltraeth Marsh has been acquired by RSPB Cymru with the intention of raising water levels and re-creating reedbeds and suitable habitats for breeding, wintering and migrating birds, especially bitterns. At Borth Bog, drainage reclamation works continued post-war, reducing and damaging the bog. However, this is now a Ramsar site, being the largest expanse of primary raised mire in the lowlands of Britain, and is part of the Dyfi Estuary National Nature Reserve.

The key addition from 1939 had been the wholesale application of technology and science, mediated through industrial capital and state intervention. By modern standards the wartime period could actually be viewed as an orgy of environmental destruction in which everything was subsidiary to ultimate wartime success, and it would be impossible, a-historic and even churlish to attempt some kind of cost–benefit analysis which weighed environmental destruction, as seen from the twenty-first century, against wartime victory as an immediate concern in the 1940s. However, a study of the impact of the war on the environment of the South Pacific notes the direct and unmeasured impacts on land, soils, vegetation, waters, as well as racial tensions brought to a head by the demands for food and supplies, and the realization of a cash economy as wartime momentum became post-war 'development'. Any concerns over environmental issues were confined to immediate use rights rather than longer-term resource sustainability (Bennett 2001, 2004). If one substitutes social tensions for the racial tensions, the situation in Britain was essentially similar.

The CWAECs achieved their goals of feeding the nation and releasing shipping from food imports, and were warmly appreciated for their efforts. But in that achievement lay the necessary conditions for the changes, that affected so much of the face of rural Britain during the following 40 years. Here, then, we see the onset of the productivist era, with fens drained and moorland reclaimed, but always subject to the particularities of their environments and the constitution and social context of the agencies involved – the CWAECs, with their sweeping powers that were only finally taken away by the Agriculture Act 1957. By that date the most compressed yet dramatic agricultural revolution in British history was well under way. It would be another 30 years before the vestiges of wartime farming policy gave way and the marginal environments could once again, in Bloom's words, be relinquished.

Notes

1 In this historical chapter, acres and miles are employed rather than hectares and kilometres. 1 acre = 0.4047 ha; 1 mile = 1.609 km; 1 ha = 2.47 acres; 1 km = 0.6214 mile.
2 *Farmers Weekly*, 22 December 1939, 8.
3 The National Archives (TNA): MAF 39/228; Parliamentary Debates (PD) (Commons) 368 21 January 1941, 66.
4 www.hinchhouse.org.uk/ninth/ninth.html; TNA: MAF 39/260–1.
5 *Farmers Weekly*, 9 August 1940, 19; TNA: MAF 70/179; 80/3855.
6 TNA: MAF 39.
7 PD (Commons) 368 23 January 1941, 299; PD (Commons) 390 10 June 1943, 846–7; 24 June 1943, 1307–8; 8 July 1943, 2250–1. On fox hunting, PD (Commons) 385 10 December 1942, 1698–9. On destruction of crops, PD (Commons) 368 21 January 1941, 65.
8 TNA: MAF 80/2813; F 37/55 Friston Forest; MAF39/300; East Sussex Record Office AMS 5666/8. *Peacehaven and Newhaven Times*, 17 January 1947.
9 TNA: MAF 38/574.
10 Museum of English Rural Life (Reading) photographic collection G9/K23186. Norfolk WAEC draining Feltwell Fen, March 1941.
11 TNA: MAF 143/21.
12 TNA: MAF 70/25; MAF 80/3644; 80/4838.
13 TNA: MAF, 112/186; 70/168.
14 TNA: WO 199/803.
15 PD (Commons) 374 9 October 1941, 1210.
16 TNA: MAF 105/288.
17 TNA: MAF 105/288.
18 Royal Society for the Protection of Birds.

References

Ashby, A.W. and Evans, I.L. (1944) *The Agriculture of Wales and Monmouthshire*, Cardiff: Society of Cymrodorion and University of Wales.

Bennett, J.A. (2001) 'War, emergency and the environment: Fiji, 1939–1946', *Environment and History*, vol. 7, 255–87.

Bennett, J.A. (2004) 'Pests and disease in the Pacific War: crossing the line', in R.P. Tucker and E. Russell (eds) *Natural Enemy, Natural Ally: Toward an Environmental History of War*, Corvallis, OR: Oregon State University, pp. 217–51.

Bevan, R. (2006) *The Destruction of Memory: Architecture at War*, London: Reaktion.

Bloom, A. (1944) *The Farm in the Fen*, London: Faber & Faber.

Bowers, J.K. (1985) 'British agricultural policy since the Second World War', *Agricultural History Review*, vol. 33, 66–76.

Bury, J.P.T. (1954) 'Short notices', *English Historical Review*, vol. 69, 508–9 [review of J. Hurstfield, *The Supply of Raw Materials* (HMSO and Longmans 1953) in the official History of the Second World War, civil series].

Carr, W. and Mercer, W. (1947) 'Reclamation of Frodsham Marshes', *Journal of the Royal Agricultural Society of England*, vol. 108, 112–26.

Cornwallis, Lord (1942) 'War-time food production: the work of War Agricultural Executive Committees. Kent', *Journal of the Royal Agricultural Society of England*, vol. 103, 76–81.

Cox, P.W. (1944) 'Front-line farming: Kent's war effort', *Agriculture*, vol. 51, 118–23.

Darby, H.C. (1940a) *The Medieval Fenland*, Cambridge: Cambridge University Press.

Darby, H.C. (1940b) *The Draining of the Fens*, Cambridge: Cambridge University Press.

Darby, H.C. (1983) *The Changing Fenland*, Cambridge: Cambridge University Press.

Ellison, W. (1943) 'Experiences in land reclamation: Montgomeryshire', *Journal of the Royal Agricultural Society of England*, vol. 104, 100–11.

Ellison, W. (1946) 'The productivity of reclaimed upland areas in Montgomeryshire', *Grass and Forage Science*, vol. 1, 54–67.

Ennion, E.A.R. (1942, reprinted 1996) *Adventurers' Fen*, London: Methuen (reprinted Cambridge: Colt Books).

Evans, N., Morris, C. and Winter, M. (2002) 'Conceptualizing agriculture: a critique of post-productivism as the new orthodoxy', *Progress in Human Geography*, vol. 26, 313–32.

Farmers' Rights Association (1948) *The New Morality*, Church Stretton: FRA.

Foot, W. (1999) 'The impact of the military on the agricultural landscape of Britain in the Second World War', unpublished M.Phil. thesis, University of Sussex.

Foot, W. (2006) *Beaches, Fields, Streets and Hills: The Anti-invasion Landscapes of England, 1940*, London: Council for British Archaeology.

Foot, W. (2007) 'The impact of the military on the agricultural landscape of Britain in the Second World War', in B. Short, C. Watkins and J. Martin (eds) *The Front Line of Freedom: British Farming in the Second World War*, Exeter: British Agricultural History Society.

Friday, L. (ed.) (1997) *Wicken Fen: The Making of a Wetland Nature Reserve*, Colchester: Harley Books.

Garrad, G. (1954) *A Survey of the Agriculture of Kent*, London: Royal Agricultural Society of England.

Grafton, Duke of (1943) 'Experiences in land reclamation: Euston, Thetford', *Journal of the Royal Agricultural Society of England*, vol. 104, 85–9.

Grafton, Duke of (1947) 'Land reclamation on the Euston estate', *Journal of the Royal Agricultural Society of England*, vol. 108, 127–9.

Harvey, N. (1995) Interview 15 February, by Phil Kinsman.

Hayter Hames, G.C. (1947) 'War-time food production: the work of War Agricultural Executive Committees: Devon', *Journal of the Royal Agricultural Society of England*, vol. 108, 86–90.

Horne, F.R. (1943) 'Experiences in land reclamation: Devon', *Journal of the Royal Agricultural Society of England*, vol. 104, 90–100.

Howkins, A. (1998) 'A country at war: mass-observation and rural England, 1939–45', *Rural History*, vol. 9, 75–97.

Hurd, A. (1951) *A Farmer in Whitehall: Britain's Farming Revolution 1939–1950 and Future Prospects*, London: Country Life.

Long, W.H. (1969) *A Survey of the Agriculture of Yorkshire*, London: Royal Agricultural Society of England.

Long, W.H. and Davies, G.M. (1948) *Farm Life in a Yorkshire Dale: An Economic Study of Swaledale*, Clapham (Yorkshire): Dalesman Publishing Co.

Mercer, W.B. (1963) *A Survey of the Agriculture of Cheshire*, London: Royal Agricultural Society of England.

Ministry of Information (1945) *Land at War: The Official Story of British Farming 1939–1944*, London: His Majesty's Stationery Office.

Moore, J. (1948) *The Blue Field*, London: Collins.

Moore-Colyer, R. (2005) 'The County War Agricultural Executive Committees: the Welsh experience, 1939–1945', *Welsh History Review*, vol. 22, 558–87.

Moore-Colyer, R. (2006) 'The call to the land: British and European adult voluntary farm labour; 1939–49', *Rural History*, vol. 17, 83–101.

Murray, K.A.H. (1955) *Agriculture* (History of the Second World War: United Kingdom Civil Series), London: Her Majesty's Stationery Office and Longman.

Orwin, C.S. (1945) *Problems of the Countryside*, Cambridge: Cambridge University Press.

Pawson, H.C. (1961) *A Survey of the Agriculture of Northumberland*, London: Royal Agricultural Society of England.

Portsmouth, Earl of (1965) *A Knot of Roots*, London: Geoffrey Bles.

Rawding, C. (2007) '"Failing" farmers in south-west Lancashire: a study of farming policy during the Second World War', in B. Short, C. Watkins and J. Martin (eds) *The Front Line of Freedom: British Farming in the Second World War*, Exeter: British Agricultural History Society.

Robinson, J.M. (1989) *The Country House at War*, London: Bodley Head.

Rose, S. (2003) *Which People's War? National Identity and Citizenship in Britain 1939–1945*, Oxford: Oxford University Press.

Scott Report (1942) *Report of the Committee on Land Utilisation in Rural Areas*, London: His Majesty's Stationery Office.

Select Committee on National Expenditure, sixth report, session 1940–1 (1941). London: His Majesty's Stationery Office.

Sheail, J. and Mountford, J.O. (1984) 'Changes in the perception and impact of agricultural land-improvement: the post-war trends in the Romney Marsh', *Journal of the Royal Agricultural Society of England*, vol. 145, 43–56.

Sheridan, D. (ed.) (1990) *Wartime Women: An Anthology of Women's Wartime Writing for Mass-Observation 1937–45*, London: Heinemann.

Short, B. (2007) 'The dispossession of farmers in England and Wales during and after the

Second World War', in B. Short, C. Watkins and J. Martin (eds) *The Frontline of Freedom: British Farming in the Second World War*, Exeter: British Agricultural History Society.

Short, B., Watkins, C., Foot, W. and Kinsman, P. (2000) *The National Farm Survey 1941–1943: State Surveillance and the Countryside in England and Wales in the Second World War*, Wallingford: CABI.

Stapledon, R.G. (1943) *The Way of the Land*, London: Faber & Faber.

Trist, P. (1971) *A Survey of the Agriculture of Suffolk*, London: Royal Agricultural Society of England.

Walford, N. (2003) 'Productivism is allegedly dead, long live productivism: evidence of continued productivist attitudes and decision-making in South-East England', *Journal of Rural Studies*, vol. 19, 491–502.

Ward, S. (1988) *War in the Countryside 1939–1945*, Newton Abbot: David & Charles.

Wentworth Day, J. (1943) *Farming Adventure: A Thousand Miles through England on a Horse*, London: Harrap.

West, W.J. (1941) 'Experiences in land reclamation: Dolfor Hill reclamation scheme', *Journal of the Royal Agricultural Society of England*, vol. 102, 98–104.

Whetham, E. (1952) *British Farming 1939–49*, London: Thomas Nelson.

Wilson, A.G. (2001) 'From productivism to post-productivism ... and back again? Exploring the (un)changed natural and mental landscapes of European agriculture', *Transactions of the Institute of British Geographers*, vol. 26, 77–102.

Woods, M. (2005) *Contesting Rurality: Politics in the British Countryside*, Aldershot: Ashgate.

3 Denial and diversity

Some reflections on emerging
agri-food geographies and rural
development in the United Kingdom

Terry Marsden

Introduction: agricultures in denial?

The last two decades of the twentieth century were a period of considerable flux in terms of European and particularly British agriculture and rural development. While the term 'crisis' is much overused, it is now clear that the 1980s and 1990s were a period when agriculture's position in society became questioned, at the same time as the British national state, for a variety of reasons, jettisoned its post-war role of providing a clear and nationally based strategy for its agricultural space by integrating its agricultural priorities with its wider town and country planning policies (Marsden *et al.* 1993). This process has been well documented over the past 20 years. It is common to trace this transition in relation to the changing nature of the state's role with respect to agriculture. This is to be expected in that in post-war years and more recently, the (multilevel) state has played a key direct (through subsidy and market regulation) and indirect role (through research and development and fiscal policy) in shaping British agriculture.

However, as I and colleagues wrote in 1986, in addressing the need for a revised and more geographically sensitive political economy of agriculture of advanced societies, it is necessary to conceptualise critically the complexity of state actions at all levels of inquiry, at the same time as understanding more broadly the uneven development of agriculture, and incorporating local and historical specificity within broader analytical frameworks (Marsden *et al.* 1986). The challenge was to understand the dynamic diversity of agricultural structures and practices rather than be content with structural dualisms (see Munton and Marsden 1991).

Since that time, many researchers have argued that a major pathway which British agriculture has followed has been one which broadly accords with a transition from a productionist to a 'post-productionist' regime or mode of regulation (Wilson 2001; Lobley and Potter 2004). In this chapter I want to readdress some of these assumptions by reconsidering the wider role of the state at the same time as continuing to problematise the complex specificity and differentiation of agriculture and rural development. I will argue that in the first decade of the twenty-first century we witness the beginnings of what some call a 'new rural development paradigm', which is influencing policy, research and practices on the ground. This requires a much more critical appreciation and redefinition of the State's (often contradictory) role in agriculture, at the same time as calling for a much-improved understanding of agricultural uneven development and 'local specificity'. While this work had its origins in the 1980s, it tended to get overtaken by both an academic and a policy-led paradigm of post-productionism which increasingly rendered agriculture a marginal pastime in the context of the growth of the 'consumption countryside' (see Department of Environment, Transport and the Regions/Ministry of Agriculture, Fisheries and Food 2000). This marginalisation and denial of agriculture is very much alive in academic and policy debates in the United Kingdom today, and it is important to critically consider some of its assumptions.

First, it rests, of course, on the full-scale critique that many of us have been party to with regard to the historical exceptionalism that the state has provided for its agriculture, and especially its farmers. Taking a broader rural development or regional development perspective, it is argued by many that agricultural support mechanisms (even the recently reformed Pillars 1 and 2 of the European Union's Common Agricultural Policy (CAP)) are geared far too inertly to the protection of the farm population, and therefore highly restrictive in their overall rural development benefits (see Lowe 2006; Ward 2006). Hence, it is assumed that we need a post-agricultural rural development policy, rather than one that still supports farmers implicitly, in terms of either agriculture or agricultural diversification.

Second, such a position assumes and generally tends to ignore one of the key findings elaborated in many of our earlier papers concerning agricultural restructuring in the United Kingdom. That is the highly adaptable, contingent and innovative nature of the farm population, not simply in terms of political mobilisation through union activities, but more profoundly through abilities to develop survival and accumulation strategies, often in circumstances that are highly threatening to their livelihoods (see Lobley and Potter 2004). Indeed conceptualisations of farmers either as individualised neo-classical price-sensitive actors or, more recently, as inert state-supported policy dupes tend to obscure the real potential of farm-based rural development. For instance, both these assumptions are alive in the UK Government's *Policy Commission on the Future of Farming and Food: A Sustainable Future* (2002), which urged farmers to compete on quality rather than price, but to do so at the same time as encouraging cost-reduction strategies across the food chain as a whole.

In this chapter I want to explore critically these assumptions in the context of a process of national, regional and local differentiation in the ways in which farm-based strategies are developing (not only in the United Kingdom but also across the European Union). This supports, I will argue, the need to escape the conceptual culs-de-sac of both the agri-industrial and the post-productivist models (which, by definition, tend to render agriculture marginal). At the same time, there is a need to begin to develop and reconstruct a more agrarian-based model of sustainable rural development that attends to the historical and innovative sociological and geographical fabric and vitality of farming.

This task is addressed by using three interlinked levels of analysis. First, at a macro level of conceptualisation I outline three prevailing and competing models of agricultural and rural development that are influencing the differentiation and uneven development of rural space. Second, I will look critically at the evidence from writers in the United Kingdom as to the nature of this process of competitive uneven development, setting out some of its key features and parameters. Third, with reference to the south-west of England, I will give examples of new forms of agrarian-based rural development that are taking hold. The evidence suggests the need for more relational and competitive parameters in assessing these new rural development processes.

Three competing models of rural development: rural space as contested space

Over the past five years I have been attempting to understand and outline the evolution of three quite distinct and competing models of rural development: the agri-industrial, the post-productivist and the rural development models. These have been defined both as heuristic devices and as major scientific thrusts that compete in and across different rural spaces to different degrees. In synthesising the contemporary writings on rural development, and in attempting to view the major forms and clusters of innovation and regulation (both technical and organisational) that are affecting rural space, we see these three models, and especially the arenas of contest and competition between them, as key drivers for rural development (Murdoch *et al.* 2003).

Table 3.1 outlines in summary the three models. The *agri-industrial model* is still seen as a major and dominant model of development across the developed and the developing world (Friedmann 2006). It is focused upon the regulation of rural space as primarily agriculturally productive space, and places emphasis upon optimum levels of production that then runs through increasingly lengthy and complex supply chains. A distinguishing feature of this model is the tendency for the value of primary production and processing at the farm and local level to decline (even with substantial state subsidies in many countries); for increasing buyer and market power to be located at the manufacturing and especially retail end of the chain; and for farmers to continue to be caught up in what many economists have termed the long-running 'cost–price squeeze'. In addition, there are two important regulatory dimensions connected to this model that involve, to varying degrees, the support of the global and national state.

These are both based upon rather contradictory positions. First, the model is supported by state activities and advocacy that supports an increasingly uneven application of the neo-liberal model. This operates north–south. It supports large subsidisation of home agricultures and agri-corporates (the United States, Europe, Japan) at the same time as projecting the liberalisation of global markets and state structures elsewhere. This is at the heart of the World Trade Organization (WTO) negotiations (see Morgan *et al.* 2006a; Higgins and Lawrence 2005). Second, as Buttel (2006) has succinctly argued with reference to the

Table 3.1 The three competing rural development dynamics

The agri-industrial dynamic	The post-productivist dynamic	The rural development dynamic
Standardised products	Rural space as consumption space	Integration
Capital intensity	Marginalisation of agriculture: declining industry	Re-embedded food supply chains
Optimum (quantitative level) of production	Share of agriculture in national income falls: 2.9% (1970), 1.0% (1998)	New policy support structures
Long or complex supply chains	Rural land as development space	Associational designs and networks
High levels of public spending	Social exclusion	Revised combinations of nature/value/region and quality
Continual development of 'technological fixes'	Public-sector services	Rural development as counter-movements
Decreasing value of primary produce and production structures	Social economy and use of the natural as an attractor in the counter-urbanisation process	Rural livelihoods/fields of activity/new institutional arrangements
Economies of scale		Agro-ecological research and development
Rural space as agricultural space		Co-evolving supply chains
Private-interest regulation (led by retailers)/ public-interest regulation: crisis management and nature management		Revised state/market/civil society/nature relations
		Evaluation paradigm for rural sustainability

United States, the model also continues to maintain both a public- and increas-ingly a private-funded research and development system which uses particular technical means (such as GM (genetic modification) or TRIPS (trade-related aspects of intellectual property rights) agreements) or 'fixes' to 'sustain the unsustainable'. This model is also based more recently on the steady internation-alisation and integration of the retailing, processing and input manufacturing sectors (Hendrickson and Heffernan 2002; Wrigley 2000).

Of course, the agri-industrial model, especially as exhibited in its more extreme intensified and monocultural forms, has been challenged significantly both by a now ingrained 'post-productivist model' and recently, on a global as well as a European level, by a more radical sustainable rural development model. Analytically, it is important to see the contested emergence of these alternatives as, in part, *reactions to the onset of considerable crisis tendencies inherent in the agri-industrial model.* Post-productivism had its origins in Europe during the mid-1980s, when a raft of financial, environmental/health and political crises besieged the sector. These have been well documented (see Lowe *et al.* 1997). They coincided, in many rural areas in the advanced world, with the further recognition of the social transformation of the countryside associated with the influx of ex-urban middle-class residents (Marsden *et al.* 1993; Murdoch *et al.* 2003). This raised the political pressure for reform of agricultural productivism, both from recently arrived local rural residents and from the gamut of environmental and food-related pressure groups. Some 25 years later it is important to reflect critically and position this process of 'post-productionism' in its specific regulatory and legitimatory context.

As Table 3.1 outlines, it has led to a particular model of the countryside which has attempted, through vast and complex regulatory instruments, first to protect larger and larger tracts of rural land from the ravages of agri-industrial-ism, and second, to attempt not to dismantle radically, but rather to confine its environmental externalities within specific boundaries and complex regulatory arrangements. This has been part of what we might call, in the European context at least, *the rise of the hygienic/bureaucratic state*, whereby, often through the hasty assembling of rules for food hygiene, safety, environmental protection or pollution control, restrictions and boundaries have been placed around 'the dirty business of industrial and intensive farming'. Such a model highlights, also, the continuing decline (in conventional GDP terms) of the economic significance of the farming sector (for instance, in the United Kingdom, agriculture comprises only 1.8 per cent of the national labour force and 0.8 per cent of gross value added (GVA)); and points to the rise of the 'consumption countryside' as now the dominant social and regulatory trend in many rural regions (Murdoch *et al.* 2003; Ward 2006; Lowe 2006; Potter 2004). Under this regulatory model, the re-emergence of environmental and rural planning, coupled with EU and national food regulation, are now seen as a key tool in developing the multifunc-tional use of rural land.[1]

It is important to recognise two important and distinguishable features of the past quarter of a century of post-productivism, at least in the European context.

First, except in the marginal cases where policy instruments have specifically enforced de-intensification programmes on farms (on livestock stocking densities, for instance), the model has been reluctant to intervene directly in the 'growth machine' and 'technological treadmill' processes which are such a central dynamic of the agri-industrial model. Rather, the recourse to new environmental planning tools has served to spatially demarcate the agri-industrial processes in ways that have produced 'buffer zones' between these and the newly designated agri-environmental spaces. (See, for instance, Van der Ploeg (2003) for a discussion of the process in the Netherlands.) It could also be argued that the recent European responses to the onset of GM through the need to develop 'coexistence schemes' between GM and non-GM land uses are the most recent manifestation of such spatialised forms of regulation. In short, then, the post-productivist model, while clearly a response to the crisis of agri-industrialism, is indeed only a palliative for it. It does not radically alter its major tenets; rather, it attempts to ring-fence and spatially manage them such that boundaries and regulations can be established between parcels of productivist and the post-productivist countryside.

The second distinguishable feature of the post-productivism model concerns the additional regulatory burdens and costs that its instruments of ostensible 'protection' have had on the local rural farm and processing sector. In short, the 'technological treadmill' of agri-industrialism (Cochrane 1979) has been combined with the 'regulatory treadmill' of post-productivism (Ward 1993; Morgan *et al.* 2006a). At the local rural and farm level this has had the effect of further deepening the traditional problems of the 'cost–price' squeeze, and further led to the decline in the number of farms, slaughterhouses and other agri-food small and medium-sized enterprises (SMEs), thus reinforcing the continuing decline in rural economic and social infrastructure. It is clear, therefore, that much of the heralded advancement of EU environmental and agricultural policies as a unique form of ecological modernisation (Frouws and Mol 1999) has also led to the driving out of production and processing of much potential rural development infrastructure. Examples of this occurring have been documented not only in the advanced countries of North-Western Europe, but also in Poland and Slovenia.

The post-productivist model (as advocated by scholars, planners and policy-makers) has, therefore, by accommodating rather than radically challenging the agri-industrial model on the one hand, and directing much of the cost and bureaucratic burden of environmental and hygiene regulation down to the rural entrepreneur or farmer on the other, *tended to reinforce its own major thesis*. That is, that agriculture, at local and regional levels, continues to be a declining industry in terms of employment, cultural capital and rural development potential. It is interesting, then, that the scientific bases of both the agri-industrial model and the post-productivist model tend to assume the declining significance in farms and local land-based businesses, either intentionally or unintentionally (Potter 2004; Ward 2006). Both assume that 'market mechanisms', often reinforced by state policies, will mean that the decline in farms and other SMEs will be inevitable. In this sense, these models are significantly influential in marginalising local forms of rural development.

Since the 1990s a third, more fledgling sustainable rural development model has been emerging, at least in a European context. While building upon the multifunctional assumptions initiated in the post-productivist model, but also adapting many of the agro-ecological principles developed in southern countries (Guzman and Martinez-Alier 2006), this model attempts to reintegrate agriculture as a multifunctional set of practices back into the rural economy. These perspectives can hold the potential to enhance the interrelationships between farms and people both in rural and between rural and urban areas. While accepting the realities of much of the 'consumption countryside' in a European context, then, the new rural development paradigm *distinguishes itself from post-productivism in the emphasis and autonomy it promotes for a reconstituted agricultural and land-based rural sector* (van der Ploeg 2000; Knickel and Renting 2000). This is seen as a potentially new driving force for rural development (O'Conner *et al.* 2006), and involves what a group of scholars refer to as broadening, deepening and regrounding activities (see Figure 3.1).

Recent studies in Europe estimate that at least 50 per cent of all farmers are actively engaged in one or other of these new rural development practices. This finding reasserts the socio-environmental role of agriculture and other land-based activities as a major agent in sustaining rural economies and cultures.[2] It reconnects a renewed priority of agricultural production to the wider markets and social innovations and possibilities (such as aspects of retro-innovation of agricultural practices (Stuiver 2006), local embeddedness (Sonnino, in press) and new forms of ecological entrepreneurship (Marsden and Smith 2005)). In contrast with the other models outlined here, which tend to assume the atomistic

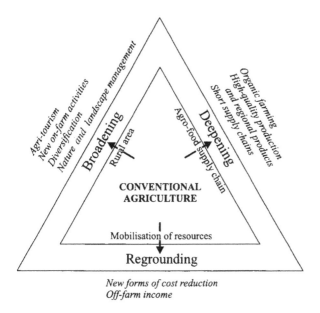

Figure 3.1 Structure of rural development at farm enterprise level.

nature of farms, and the neo-classical or state-led predisposition of their farmers, the rural development model suggests a re-created potential for symbiotic inter-connectedness between networks of farms and farmers in the same locale. This is based upon assumptions of the associational economics and institutional economics literature (see Morgan *et al.* 2006).

In this context, agriculture acquires a more comprehensive meaning and displays higher integrative potential in that it is increasingly recognised as a central feature in delivering real rural sustainable development benefits to both rural and urban populations. Moreover, this does not necessarily rely upon significant amounts of production subsidy, but rather, as we shall see, can engage the state in new and innovative ways (for example, as in aspects of marketing and procurement policies). To do this, however, it is necessary to create *a radical rupture with the agri-industrial processes mentioned above. Agriculture must, in a variety of ways, attempt to find new political, social and ecological platforms and spaces to distinguish itself from the conventional modernisation processes that tend to continue to devalue its base.* Empirically, we see many examples of this across Europe in what has been termed the 'relocalisation' of agri-food (Marsden and Murdoch 2006). While these re-embedding tendencies have proliferated in some countries (such as Italy and France) more than others (such as the United Kingdom and the Netherlands), it is now clear that even in those countries where the agri-industrial model historically has had the most profound impact, new 'socio-technical niches' and spatially embedded networks and structures have developed as a significant response to it. I will explore these processes in more detail below.

This is not just associated with the rise of organic food production and consumption in Europe (which has indeed been significant). It is also associated with, first, the rise of both locality-based and new locally-based short food supply chains (Renting *et al.* 2004), and second, the different responses and ruptures from the conventional system brought about by its periodic and severe crises tendencies. For instance, Sonnino and Marsden (2006) have analysed the relocalisation process in south-west England. Here, new and more autonomous relocalised networks of locally and regionally based beef, dairy and cheese production have developed, partly as some producers have attempted to create a rupture with the past conventional systems (for instance, the bovine spongiform encephalitis (BSE) and the foot-and-mouth disease (FMD) crises). Rather than continue to struggle under the continual pressures of the (agri-industrial) conventional 'cost–price squeeze' on the one hand, and/or face the increasingly burdensome regulatory costs of the (post-productivist) hygienic-bureaucratic state on the other, these new networks have managed to detach themselves from 'lock-in effects' by developing new sets of relationships both between themselves and with downstream buyers (such as retailers and caterers). It is also clear that these new networks begin to return considerable economic value added for the local rural economy.

There are many critics of the new rural development paradigm, as we might expect. Many agricultural economists see it as really only another novel form of 'niche production' in the global and World Trade Organization (WTO) context of liberalisation and continued agri-industrialism, an archipelago within the sea of corporate-inspired cost reduction. Some question its more widespread benefits both socially and economically, arguing that these new networks are restricted, and indeed display quite neo-liberal tendencies of their own (Du Puis and Goodman 2005), or that they can also reflect the more sinister features of 'defensive localism' and protected markets (Winter 2003). More imaginatively, some suggest that this movement begins to represent a sort of postmodern *re-peasantisation*, whereby new networks of landholders begin to develop more flexible systems of autonomous and endogenous rural development. It has been clear, however, since at least around 2000, that we can witness the arrival of this new model, albeit a model that struggles in a competitive context of the two more powerful and state-supported models outlined earlier in the chapter.

The differential role of the state

One of the distinguishing features of the three competing models of agri-food and rural development outlined here relates to the differing roles that the state plays in each case. The rural development model usually has to rely on what some scholars are calling the competitive 'project-state' – that is, funding which is largely EU based, and allocated on the basis of competitive bidding and several conditions (such as partnership working). These funding mechanisms, such as LEADER (the EU rural development programme) and regional develop-ment structures funding, have clearly been important in stimulating the rural development paradigm across Europe. Also, the protected area-of-origin desig-nations (PGI and PDO, which denote EU juridical protection of quality produc-tion within specific areas) have been extensively if unevenly used across member states (see Renting *et al.* 2004). The 'project state' approach may well continue in this model, with the arrival of the 'second pillar' of the CAP, whereby funding can be allocated for farm-based diversification and multifunc-tional ventures. It is as yet unclear whether reform of Pillar 1 of the CAP will further stimulate these trends.

It is important to recognise, however, that the increasingly hybrid nature (that is, public and private systems of governance) of regulation which now domi-nates both the agri-industrial and post-productivist models can have the effect of continuing to marginalise the rural development model. For instance, the Euro-pean model of food safety regulation is dominated increasingly by 'scientific-based' risk assessment and management principles, at the same time as allowing the corporate retailer sector to create global and privately organised 'traceabil-ity' systems on behalf of the more risk-sensitive food consumer (Smith *et al.* 2004). These can stifle the necessary 'room for manoeuvre' upon which the rural development model relies.

The state thus plays key roles in all three models, and in this sense one might argue that it holds increasingly contradictory tendencies in attempting to deal with food regulation, agricultural production support, environmental protection and sustainable rural development in different ways in different rural spaces. For instance, the growing power of consumer lobbies in Brussels, together with the articulation of European consumer concerns in European Parliament debates, has tended to heighten the significance of food safety, health and quality policy over the narrower agricultural reform agenda. The rural development model would seem to rely more on the local and the regional state, for it is at this level that effective institutional mechanisms can be put in place to protect and foster local forms of innovation in the agri-food sector, for instance.

Parallel models of rural governance and uneven development of agri-food: the global context

It is clear that these parallel and competing rural and agri-food development models are operating at different spatial and institutional scales, both in Europe, where they have most clearly been applied, and elsewhere. While they operate at a relatively high policy and institutional level (as in the case of global and regional policy institutions, such as the WTO, World Bank or the European Union) in both scientific and policy discourses, it is also important to recognise that they also shape the local and regional levels of rural development. Research work conducted in China and Brazil, for instance, indicates that certain agricultural regions are experiencing both the profound influences of the agri-industrial model of regulation, through, for instance, the use of privatised systems of quality and safety regulation, on the one hand, and more locally embedded attempts by groups of smaller producers and state bodies to develop their own systems of more sustainable and endogenous forms of rural development which lie outside these often export-oriented and 'marketised' systems of regulation (Van der Grijp *et al.* 2005).

One paradox of these new private-interest food governance systems is that, while they are designed by international retailers to create greater standardisation of product and assurance for the European consumer, they are also likely to create new forms of uneven development between local producers and workers (Hatannaka *et al.* 2006). Also, this process of uneven development can become exacerbated the more such regions come to rely upon export markets, a process clearly apparent in the São Francisco Valley in north-east Brazil since about 1990.

Increasingly, the main state agencies in the region are also realising the need for more integration with the alternative rain-fed, non-irrigated and non-export-based sectors. As in parts of rural China, local agricultural ministry officials in the Brazilian agricultural research corporation (EMBRAPA) are attempting to develop more endogenous strategies, such as by stimulating demand for local goat and sheep meats, and developing organic fertiliser markets. These are alternative, rural development attempts to obtain improved, 'win–win' solutions between continued market development, local development and sustainability in a competitive context that has hitherto prioritised meeting the demands of externally controlled, export-based supply chains. These alternative concerns become a difficult balancing act for state officials in the region, with the large export enterprises placing considerable pressure on state authorities to provide them with research and development expertise geared to increasing 'quality' export markets. Hence, there is a critical tension between the forces of agri-industrial standardisation and quality control, and promoting more rural diversity and development in the region.

In this context, many new but relatively small 'alternative' projects are being mounted. Organic production provides an emergent but largely unexplored avenue. One large farmer in the region is unique in that he has converted the entire farm to organic production, while others are experimenting with parts of their holdings. He holds full certification from the Brazilian organic certifiers, and exports all his mangoes to the United Kingdom and the Netherlands. He argues that he can get higher premiums in export markets and that his production costs are lower given the savings from not using artificial fertilisers and pesticides. However, the marketing and distribution chains are less reliable and far less institutionalised than in the established conventional export sectors.

In the case of many newly industrialising and developing countries and regions we see that whereas the post-productivist model may be far less applicable (partly as a result of more sparsity of urban wealth and environmental activism), the agri-industrial model and the alternative rural development practices are developing alongside one another and often competing for scarce knowledge, finance and market-based resources. In developing countries, owing to a lack of robust institutional support, resources and know-how, and the global and national pressure to export in what are supposed to be freer markets, it may be far more difficult to gain widespread support for sustainable rural development. However, it is not impossible. In Europe, despite the continuance of agri-industrial thinking and policy, and the onset of a raft of state-led post-productivist policy and reform (such as the slow reform of the CAP from production subsidies to conditional environmental management payments to many farmers), it is clear that there is evidence that spaces have opened up for alternative and agricultural or agri-food-based forms of rural development (van der Ploeg 2000; Marsden and Murdoch 2006).

It is important to examine critically the contemporary European and UK experiences in this global context, given that one feature of the past 25 years has been a growing integration with these wider global processes of agrarian and agri-food change, through both the rapid development of transitional retailer-led supply chains and the broader reshaping of trade and production systems. Moreover, it is unlikely that the pressure upon Europe to reduce its agricultural protection will weaken.

The contested nature of agricultural change in the United Kingdom: to what extent are we seeing the arrival of a sustainable rural development model in the United Kingdom?

The above conceptualisations, which suggest the highly contested nature of rural change in different regions of the world, are therefore also very relevant when we consider contemporary UK rural space. Indeed, over the past two decades the United Kingdom has become more highly influenced by the 'Europeanisation' of both markets and regulation (see Smith *et al.* 2004) and the globalisation of food chains (Van der Grijp *et al.* 2005). At the same time, UK agriculture has integrated itself further underneath both productivist and post-productivist agricultural and environmental policies. In short, the above analysis suggests that we need to explore the contested nature of this 'battleground' of rival paradigms in and through UK agricultural and rural space, rather than rely upon more generalised 'from–to' transitions, such as from productionism to post-productionism (Walford 2003a, b).

Hence, we can argue that there are at least three sets of dynamics (agri-industrial, post-productionist and rural development) occurring *at the same time in UK rural spaces*, and at local and regional levels it should not be surprising to find that elements of all three are being expressed in complex, sometimes contradictory, and more nuanced ways. It is a necessary analytical function of the conceptualisations above to assist in understanding and seeing the reasons and influences upon rural and agricultural differentiation. In short, we need better theory as well as empirical evidence to understand rural differentiation (that is, to theorise about why things are different in and through space), and this needs to marry broader conceptualisations with highly differentiated and complex empirical realities.

Such arguments of what we might term intense competitive differentiation are reflected, but also somewhat obscured, in both the recent policy and academic literature concerning agricultural restructuring in the United Kingdom. Writers such as Ilbery and Watts (2004), Walford (2005) and Lobley and Potter (2004) point to the geographical complexity of UK agricultural restructuring and differentiation, as well as to the over-generalised productivist–post-productivist transition. Ilbery and Watts (2004: 1–2), for instance, argue that

> [t]he idea of a transition is important because the second phase of agricultural change has not replaced the first. There is, for example, evidence that farmers still think in a productivist manner (Winter and Gaskell 1998; Walford 2003). The two phases of agricultural change *thus co-exist in what is being described as a multi-functional agricultural system* (Potter and Burney 2002; my emphasis).

As a result, such scholars often see these as contradictory trends in the official data. Lobley and Potter (2004: 508) also state, in their more comprehensive study of agricultural restructuring across six regions of England, that

> it is perhaps inevitable that these results tell a more complicated and geographically nuanced story of agricultural change than much public commentary would lead one to expect.... A well-defined core of professional and professionalizing farmers emerges from the survey, confident in their ability to fine-tune their businesses to changing market and retail

demands and for whom 'post-productivism' seems a very remote concept. For a significant minority, however, the economic centrality of agriculture at a household level looks set to further decline.

Taking empirical cross-sections of data across different regions of the United Kingdom thus tends to reinforce the view that there is increasing differentiation both between and within regions, and that this suggests different farm-level strategies of intensive engagement and concentrated landholding or disengagement from conventional farming and commodity production (Walford 2005). Such analyses, while empirically rigorous, have largely failed to contribute more theoretically or conceptually beyond the point that the productivist–post-productivist dichotomy is much more complicated. In addition, the continued inadequacy of official farm statistics (still largely commodity based) downplays farm-based structural change regarding this 'nuanced story'. In consequence, academic and policy debates concerning the new 'farm structures question' are hampered, and real evidence on how conventional and alternative strategies are being actively constructed and realised is not available. All this severely hampers understanding of the ways in which a more differentiated agriculture could contribute, or indeed does contribute, to a more sustainable rural development model.[3]

It follows from this point that in order to address the extent to which a more sustainable rural development model for UK rural space can or could really take hold, it is necessary to explore its emergence in the highly competitive and regulatory context of continued agri-industrialism *and* post-productivism. Hence, seeking to identify the emergence of case examples of rural development on the ground alone does not do justice to our understanding of competitive differentiation in the UK agricultural sector. So, what are some of the key factors in this competitive process in the United Kingdom?

The continuing deep currents of agri-industrialism and post-productivism

It is clear that over the past 20 years, conventional agriculture has increasingly been practised on a smaller number of larger farms. However, this does not suggest the complete eclipse of agri-industrialism in the United Kingdom, since the number of conventional producers may have fallen, and those practising some form of diversification may have increased. Commercial dairying and potato-growing are good examples of sectors that have experienced declining numbers of producers (for example, falling in England and Wales from 71,800 in 1998 to 58,700 in 2003; and from 18,000 to 12,200 respectively). In this sense, the cost–price squeeze and the operation of a combined technological and regulatory treadmill is continuing to reduce the number of farmers and farm businesses in the United Kingdom as a whole. Figures 3.1 and 3.2 indicate some of the multiple pathways that have been identified from empirical evidence contained in various sources. For those committed to mainstream agriculture, the

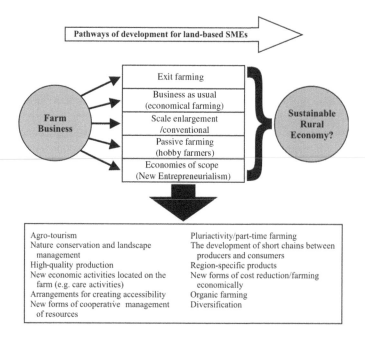

Figure 3.2 Potential pathways for farm-based businesses.

emphasis has continued to be on enlarging the *economies of scale* rather than those of the *economies of scope*; although it is argued now that 46 per cent of farms in England had some form of diversified activity by 2004–5, with 60 per cent of these having an output value of less than £10,000 per year, but with 25 per cent of such businesses having a diversified income which exceeds that gained from conventional farming (Department of Environment, Food and Rural Affairs 2006) (Table 3.2). For the conventional farmers – and there is a lack of really detailed empirical work on this process – the search has been on to take over adjoining farm land and buildings and to consolidate larger farm businesses around the core business. This process is continuing despite reform of the CAP in 2003 (Walford 2003a, b; Lobley *et al.* 2002).

It is also important to recognise that over the past two decades the economic power of the farm sector has declined vis-à-vis other actors in the food chain. Figure 3.3 shows that by 2005 there were just over 300,000 farmers producing a gross value added of £5.2 billion, and creating 541,000 jobs (increasingly part-time). This was only a quarter of the value of food imports (£21.9 billion) or the gross value added of the grocery retailer (£20.4 billion) or the catering sectors (£21.8 billion). Moreover, compared with 1998, the farmers are estimated to have received 25 per cent less in 2004 for their basket of UK food (Department of Environment, Food and Rural Affairs 2004–5).

These trends indicate the increasing dependence of the UK agri-food sector

Table 3.2 Value of output from diversified enterprises, England, 2004–5

	No. of farms	Percentage of all farms	Output (£ m)	Ave. output (£/farm)	Ave. margin (£/farm)	Margin (%)
Farm business output (incl. diversification)	60,800	100	11,000	165,400	27,100	16
Diversified enterprises	27,800	46	510	18,500	10,900	59
of which:						
Letting buildings for non-farming use	21,700	36	270	12,400	9,500	77
Processing/retailing of farm produce	4,300	7	110	26,000	11,300	43
Sport and recreation	4,200	7	26	6,100	3,600	59
Tourist accommodation and catering	2,600	4	26	10,000	5,300	53
Other diversified activity	2,300	4	79	33,800	7,600	22

Source: Farm Business Survey.

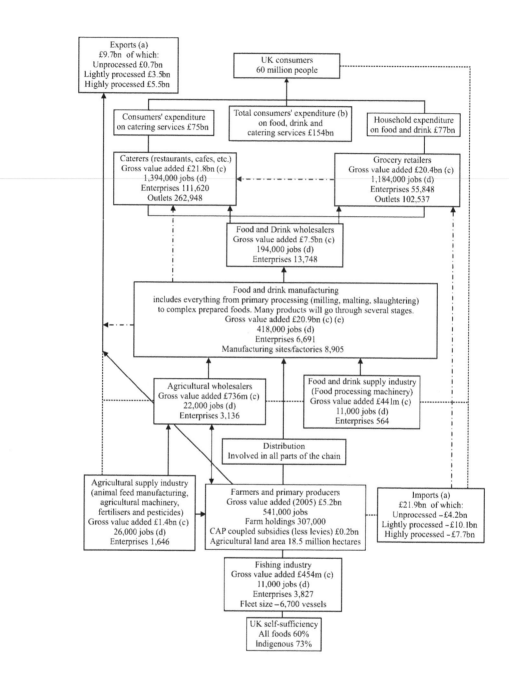

Exports (a)
£9.7bn of which:
Unprocessed £0.7bn
Lightly processed £3.5bn
Highly processed £5.5bn

UK consumers
60 million people

Consumers' expenditure
on catering services £75bn

Total consumers' expenditure (b)
on food, drink and
catering services £154bn

Household expenditure
on food and drink £77bn

Caterers (restaurants, cafes, etc.)
Gross value added £21.8bn (c)
1,394,000 jobs (d)
Enterprises 111,620
Outlets 262,948

Grocery retailers
Gross value added £20.4bn (c)
1,184,000 jobs (d)
Enterprises 55,848
Outlets 102,537

Food and Drink wholesalers
Gross value added £7.5bn (c)
194,000 jobs (d)
Enterprises 13,748

Food and drink manufacturing
includes everything from primary processing (milling, malting, slaughtering)
to complex prepared foods. Many products will go through several stages.
Gross value added £20.9bn (c) (e)
418,000 jobs (d)
Enterprises 6,691
Manufacturing sites/factories 8,905

Agricultural wholesalers
Gross value added £736m (c)
22,000 jobs (d)
Enterprises 3,136

Food and drink supply industry
(Food processing machinery)
Gross value added £441m (c)
11,000 jobs (d)
Enterprises 564

Distribution
Involved in all parts of the chain

Agricultural supply industry
(animal feed manufacturing,
agricultural machinery,
fertilisers and pesticides)
Gross value added £1.4bn (c)
26,000 jobs (d)
Enterprises 1,646

Farmers and primary producers
Gross value added (2005) £5.2bn
541,000 jobs
Farm holdings 307,000
CAP coupled subsidies (less levies) £0.2bn
Agricultural land area 18.5 million hectares

Imports (a)
£21.9bn of which:
Unprocessed –£4.2bn
Lightly processed –£10.1bn
Highly processed –£7.7bn

Fishing industry
Gross value added £454m (c)
11,000 jobs (d)
Enterprises 3,827
Fleet size –6,700 vessels

UK self-sufficiency
All foods 60%
Indigenous 73%

on imported products on the one hand, and declining value added at the farm level on the other. It suggests that reduction in the number of conventional farm businesses is likely to continue, with smaller farmers facing 'horizontal pressure' to sell up to their scale-enhancing neighbours on the one hand, and 'vertical pressure' from both imports and retailers to suppress farm gate prices for conventional products on the other. We can also see from these trends how, even if we characterise the domestic agricultural scene as post-productionist, this is increasingly based upon a prevalent productionist and agri-industrial intensive model in the procurement of imports from overseas. Under these conditions we can see that the retailer-led cost–price squeeze, coupled with cheaper imports, will continue either to drive farmers down a path of scale enlargement and concentration, or to create viable but risky exit strategies from these conditions. We should not, however, underestimate the abilities of existing farm families to adapt and adjust to these pressures, as 'persistence' has been a major feature of the sociology of modern agriculture. Indeed, as Lobley *et al.* (2002: 2) argue:

> Many English farm families are long established. Our surveys found that almost one-third had been farming in the same area for more than a century. Only 8 per cent were new entrants in the sense that they were the first generation of their family to farm and had not previously farmed elsewhere. A high proportion expects to remain in farming despite economic recession, BSE and FMD and a prime motivation of their restructuring activities is to achieve this goal. Taken together, our findings suggest that the agricultural and agri-environmental assets of rural England are in (and will remain in) relatively few, unchanging hands for the foreseeable future.

While this may be the case in aggregate terms, in terms of their distribution there are still likely to be fewer, but larger, *conventional farmers* and a continued growth in those who adapt through different forms of diversification or further intensification. This is again especially the case in the UK dairy sector, where the abolition of former cooperative arrangements through the particular implementation of government competition policy, in addition to retail pressure on farm gate prices, continues to make increases in scale and intensity of production a necessity. Supply-chain pressures are likely to be exacerbated by the

---

Figure 3.3 The UK food chain (DEFRA 2004–5).

Notes

a Overseas trade data are provisional for the full year 2004 from HM Customs and Excise.

b Consumers' expenditure, properly known as household final consumption expenditure, is a provisional estimate by DEFRA for 2005 calculated at current prices.

c Gross value added figures are provisional data from the Office of National Statistics for 2004 calculated at basic prices (market prices less taxes plus subsidies).

d Employee data are for Q3 2005 from the Office for National Statistics.

e Gross value added for food manufacturing does not include farm animal feed, which is included in agricultural supply industry.

2005 CAP reforms, whereby dairy premiums have been decoupled and incorporated into the single farm payment system. This could lead to a major shake-out of dairy farms, estimated from a current base of 25,000 to 15,000 by 2015 (House of Commons 2004; Morgan *et al.* 2006a) and creating more reliance upon imports.

Rogaly (2006: 6) also documents this intensification process with reference to the British horticultural sector. One response has been a rapid growth in part-time and casual migrant labour. He argues:

> Our case histories suggest that, while there has been a decline in the availability of long-term residents, including British nationals, and an increase in migrant workers willing to work in the sector, the main reason for this structure of demand *lies in the relations between growers and retailers* [emphasis added].... This has meant declining margins available for growers on each unit of output. Many producers of fresh fruit and vegetables have gone out of business, as evidenced by the shrinking and increasingly concentrated structure of the fresh fruit and vegetable sectors. Others have sought what they saw as the only viable way forward: to supply greater volumes, through intensifying production and becoming involved in the packing and primary processing not only of their own products, but also of imports.

Swimming against the tide: counter-movements and the emergence of rural development in the United Kingdom

As Figures 3.1 and 3.2 indicate, however, this is only part of the current story of the political economy of contemporary UK agriculture. Farmers are increasingly opting for different combinations of what we have previously termed *broadening, deepening and regrounding strategies*, at the same time as defining counter-strategies to those more traditionally associated with the agri-industrial model. In a recent study of these activities we produced estimates and projections of such activities for farms in England and Wales (see Tables 3.3 and 3.4). The different types of rural development activities are estimated as contributing to 32 per cent of total farm family income (in 1998), with only 31 per cent of a random sample of farms obtaining all of their farm income from farming in 2002 (see Marsden and Parrott 2006). While these farm-based activities are a more recent departure and considerably less widespread than in other EU countries, they have been growing rapidly over the past decade and look set to continue and consolidate their growth (Tables 3.3 and 3.4). The deepening and broadening activities with the strongest take-up in England and Wales are agri-tourism, short food supply chains and new on-farm activities. These also make significant contributions to the net value added (Marsden and Parrott 2006; Renting *et al.* 2004). Such new rural development processes are also highly regionally differentiated. Significant variations exist, for instance, between the West Midlands (low) and the South-West (high), and some of the reasons for

Table 3.3 Projection of impact levels for deepening, broadening and regrounding activities for the United Kingdom, 1998

Fields of activity	Number of farms	Percentage of total N	Extra NVA/FI per farm (€)	Extra NVA per FoA (million €)	Percentage of total NVA	Extra FI from RD-activities (million €)	Percentage of Total FI
Deepening							
Organic farming	1,462	0.6	16,990	25	0.2		
Quality production	3,193	1.4	16,763	54	0.5		
Short supply chains	14,713	6.3	21,609	318	3.0		
Subtotal deepening				397	3.7		
Broadening							
Agri-tourism	19,417	8.3	17,039	331	3.1		
New on-farm activity	16,201	6.9	13,021	211	2.0		
Diversification	10,740	4.6	7,150	77	0.7		
Management of nature and landscape	46,255	19.8	1,540	71	0.7		
Subtotal broadening				690	6.4		
Total D+B	100,784	43.2	10,777	1,087	10.1	642[a]	8.5
Regrounding							
Cost reduction	52,963	22.7	10,965			581	7.7
Off farm income	116,776	50.1	9,994			1,167	15.5
Subtotal regrounding						1,748	23.2
Total RD (D+B+R)						2,390	31.8
Reference data total Agriculture sector in the United Kingdom	Total N 233,150			Total NVA 10,754		Total FI[b] 7,525	

Source: Unpublished IMPACT matrix data; Marsden and Parrott (2006).

Notes

NVA, net value added; FI, farm income; FoA, Field of activity; RD, rural development; TFI, Total Farm Income

a The calculation of extra Total Farm Income for the UK assumes that 59.1% of the extra NVA derived from deepening and broadening can be considered as extra FI. This figure is based on the relationship between NVA and Total Farm Income in the UK, as published in the *Eurostat Farm Structure Survey 1997* (the most recent available year).

b Total Farm Income calculated as Farm Income from farming plus Off-Farm Income= 6,358,000,000+1,167,104,339 (Eurostat Farm Structure Survey for 1997).

Table 3.4 Potential impact of deepening, broadening and regrounding activities in the United Kingdom, 2008

Fields of activity	Number of farms	Percentage of total N	Extra NVA/FI per farm (€)	Extra NVA per FoA (million €)	Percentage of total NVA	Extra FI from RD-activities (million €)	Percentage of Total FI
Deepening							
Organic farming	8,847	4.3	16,955	150			
Quality production	5,240	2.5	16,631	87			
Short supply chains	23,965	11.6	26,342	631			
Subtotal deepening				868			
Broadening							
Agri-tourism	22,844	11.1	21,675	495			
New on-farm activity	18,745	9.1	19,736	370			
Diversification	16,661	8.1	10,177	170			
Management of nature and landscape	74,133	36.0	2,114	157			
Subtotal broadening				1,191			
Total D+B				2,060		1,218	15.1
Regrounding							
Cost reduction	52,475	25.0	12,062			621	7.7
Off-farm income	123,600	60.0	12,791			1,581	19.6
Subtotal regrounding						2,202	27.3
Total RD (D+B+R)						3,420	42.4
Reference data total Agriculture sector in the United Kingdom	Total N 206,000			Total NVA 10,969		Total FI 8,066	

Source: Marsden and Parrott (2006).

this lie in a variety of geographical factors such as tradition, tourist potential, regional state support and degree of integration into the conventional (retailer-led) sector. In short, as I will explore, these new forms of rural development rely upon a redefinition of the local and regional resource base and a different set of 'initial endowments' with regard to creating value added.

More significantly, it is important to see the arrival and development of these more 'multifunctional' farm activities as new sociological and geographical expressions of a paradigm shift away from the conventional, and indeed post-productivist, systems. In the twenty-first century they are not just about farm diversification or 'multiple-income sources', trends that have been well documented from the 1980s (Walford 2003a, b). Rather, our evidence suggests real shifts in thinking and approach by many of the adopters towards a new reintegration and a refreshed and multidimensional productionism based upon new forms of 'ecological entrepreneurship'. Above all, this is ushering in new forms of regional and intra-regional uneven development between conventional farmers and multifunctional businesses, and indeed between such multifunctional businesses themselves. As our earlier evidence suggests, the arrival of the new rural development model is far less coherent. It is highly variable in both activity and practice on the one hand, and in terms of its distinct need to be variably embedded in and through places and natures on the other, as we will conclude below.

I will conclude here by identifying some of the key revised uneven development parameters that are emerging as a result of these new rural development initiatives, drawing on recently completed research in south-west England.

Respacing agricultures in south-west England: new forms of associationalism

New forms of embeddedness embrace not only the socio-economic dimension of food production and consumption activities, but also their wider ecological and cultural context (see Barham 2003; Sage 2003). Some regions, which have lost their sense of food consumption culture (Day 2006) and where the conventional model of distanciation between production and consumption is dominant, find this process of re-embeddedness much more difficult. However, in the South-West we discovered an increasing and widespread emphasis upon local and regional food provenance. This is complemented by a vibrant tourist industry and the strategies of corporate retailers to begin to meet consumers' demands for local and regional foods (Morgan *et al.* 2006b).

These conditions have provided a strong basis for the development of new networks of producers who *regear* their businesses so as to meet value added market opportunities. Such new networks develop their own agricultural and social geographies, providing networks of farmers with new marketing outlets, based upon shared values and goals associated with the relationships between the 'bio-region', agricultural practices and the active engagement in the marketing of products. As such, the networks display a new rural development form of productionism and occupy significant tracts of rural and agricultural space. Three cases exemplify these shifts:

1   For a group of Cornish producers, the production of clotted cream both as a local product on the regional tourist market hotels, restaurants, shops and cafés, and as a wider 'locality' product on the national retailer market is providing a new platform for rural development. Based on re-creating traditional processing techniques, the Rhodda creamery is now at the centre of a farm network that controls 80 per cent of the market for Cornish clotted cream. These producers argue that Cornish 'special grass' implies that animals produce 'better milk', which then makes a product of superior quality. Thus, they use revised 'bio-regional' criteria to compete in the quality markets and to distinguish themselves in the competitive marketplace for clotted cream. In addition, this means that some of the producers in the supply network of the cream can exploit the 'economies of scope' by co-producing, for instance, raw milk, cream, yogurt and milk waste products for pig feed. As one producer commented:

> Our business is all about a balance between, say skimmed milk and clotted cream.... Because if you sell a 4-pint bottle of skimmed milk, you have got no choice but to have the cream, so you've got to sell it, you have to do something with it. What pays our bills is the combination: a 4-pint bottle of skimmed milk and a pot of cream; the combination, that's really where we get the margin ... we offer a full range of products; a lot of people like the milk, other people like the butter, others the clotted cream. It's a combination, really.

2   Steve Turton, a former farmer and family butcher from Exeter, now lies at the centre of a network of 154 local farmers and processors supplying 100 per cent traceable meats to shops, restaurants and, most significantly, Sainsbury's supermarkets. In 2001 he managed to persuade the retailer to introduce Turton's Westway Sausages in seven stores in the region, which sold 200 per cent above Sainsbury's own leading brand. By 2003, Turton was selling 196,000 cases of sausages through the supermarket, running through 15 stores. As a former master butcher, he argues that 'regionality and traceability, combined with an emphasis on quality', formed the key to success of the business and the network. Developing relationships with producers and processors who share the same values and goals has been crucial to the success and development of the network. As one key supplier argues:

> With Steve Turton we are visiting the farm, we are discussing what we are going to breed, how we are going to feed it, when we are going to produce it ... It's a partnership arrangement, I just have to fit with his philosophy ... I've got two of the best eating quality breeds, and he's looking for eating quality, so we fit.

3   This proactive form of bio-regionalism becomes formalised and embedded
     through the development of a 'quality' regional brand, as in the case of
     Turton meats, or sometimes through the more formal PDO certification. The
     emphasis always has to be on distinguishing continually the quality product
     from more conventional products and also from the growing number of
     alternative 'quality' products. This is especially the case for Cheddar
     cheese, which is produced nationally by large processors. The case of
     Cheddar illustrates how a PDO certification can position a producer's brand
     on a highly competitive market. After the abolition of the Milk Marketing
     Board, two dozen farmers in Somerset, Devon and Dorset who had
     traditionally produced Cheddar cheese established their own cooperative so
     as to compete with the emerging branded quality cheeses. This was an
     attempt by the farmers to get more involved in the marketing of their prod-
     ucts and to directly negotiate quality and price with buyers. In 2000 the
     cooperative was awarded PDO status, and at the same time it introduced its
     own brand: West Country Farmhouse Cheddar (WF). Seen as a marketing
     tool that protects the artisanal Cheddar from the competition of conven-
     tional cheeses, the PDO becomes for many producers an ideal means to
     make quality and geographical distinctions with the standardised and highly
     competitive 'block' cheese market, as well as to develop more direct links
     with major retailers and caterers. In addition to having their own farm label,
     producers and processors of WF Cheddar also combine their own individual
     producer brands (such as Dennay) under the cooperative umbrella label.
     These multiple branding strategies are managed through the cooperative,
     which operates as a gatekeeper for such branded niches, manages and regu-
     lates quality control, and negotiates with the major retailers. It achieves a
     premium price of £2,000–3,000 per ton with the supermarkets.

These examples have originated out of ruptures with the dominant and conventional food system. These ruptures were often triggered by the crises that
have beset the industry over recent decades, notably BSE and FMD. However,
they represent significant developments that in many ways lie adjacent to the
conventional systems of production. In the South-West, conventional dairy
farming is struggling, and many of our respondents were pessimistic concerning
the long-term benefits of the reforms of the CAP system (single farm payments
and Pillar 2 monies).

Conclusions: theorising difference, theorising competition

The emergence of these new, more ecologically embedded agri-food networks
has now been documented and debated in the recent economic geography literature (see Sonnino and Marsden 2006; Watts *et al.* 2005), and is connected more
broadly to the relational perspectives currently given focus in this literature (see
Bathelt 2006). What is clear is that these are based upon new sets of social and
economic relationships, not only between groups of farmers, but between

farmers, processors, retailers and, to varying degrees, different segments (local and non-local) of the consuming public. New horizontal and vertical reconnections are made between these actors and, as a result, agricultural and rural space is being remade. They thus rely upon new forms of relationalism and associationalism.

While the regional and local state may be supportive in various ways (for example, in allocating competitive rural enterprise grants or supporting the bids for PDO/PGI recognition), such networks are less reliant on state subsidies or on the bureaucratic nature of post-productionist regulatory systems. Rather, they 'carve out' newly embedded competitive spaces both despite and in the midst of these regulatory systems, and in competition with the more conventional forms of agricultural production. These new micro-geographies thus sit alongside and in competition with the conventional (agri-industrial and post-productionist) systems, thus creating a more diverse and richer patchwork in the agricultural geography of regions such as the South-West.

In summarising the burgeoning growth of these networks, we can identify at least four socio-geographic parameters that suggest the need for a much more sociological and geographically embedded appreciation of farmer-based innovation, as was argued in the Introduction (see Sylvander *et al.* 2005). First, the empirical research so far points to the need for a much deeper understanding of the *processes of resistance and rebuilding* that are experienced by landholders as they struggle to 'exit' from (the lock-in effects of) conventional systems, and create new networks and platforms of both 'quality' production and supply.

Second, it is clear, not least from the brief cases outlined above, that a critical feature of these networks becomes the abilities to negotiate, either individually or collectively, with a whole host of external agencies and actors (especially corporate retailers in persuading them to open up shelf space for their products). A significant feature of such networks, therefore, is their *attempts to create the power to negotiate* around concepts of local and regional space, branding, and shelf space, as well as price. In this context, the changing and contingent strategies of retailers is an important factor in the continued development of networks and in reflecting growing consumer demand for 'quality' and regional products.

Third, and directly related to the contingencies of consumer demand and retailer strategy, such networks, once established, have to invest continued time and resources into *creating and maintaining competitive boundaries* based upon their particular sets of ecological and bio-regional quality conventions. The competitive and entrepreneurial task becomes that of *continuing to distinguish products*, not only from the conventional sectors but also from the growing number of other quality suppliers. This is a new form of relational competition between different networks (both within and between regions) that was, and still is, significantly absent in the CAP-based support system. It relies upon the creation and articulation of knowledges and techniques associated not only with the foods themselves, but also with the particular bio-zoological-regional context in which they are produced. In this sense, many of the actors in the networks become 'geographers' themselves, revising, redefining and reshaping the agri-cultures around them.

A fourth dimension concerns the priorities that need to be placed on main-taining some *level of structured coherence within the networks themselves*. This involves key actors being able not only to maintain trust but also to demonstrate continually the economic value added from being in the network, as opposed to being outside it. As one of our case-study respondents noted, it is important for him to keep his network of suppliers 'hooked' – that is, to demonstrate continu-ally the advantages of association as opposed to individualism.

The latter part of the chapter has begun to show, but only with reference to agri-food networks, how new and competing relational agricultural geographies are emerging in the United Kingdom out of the wider global, European and national processes of agri-industrialism, post-productivism and the more recently emerging rural development models. Some scholars and commentators may still regard the agricultural sector as increasingly residual with regard to its impact on the wider rural economy. However, if we are to take rural sustainabil-ity seriously, then scholars and policy-makers will need to reintegrate these rela-tional agricultural geographies into their broader understandings and actions of rural development more generally.

Notes

1 This post-productionist stance continues to be conveyed by policy-making bodies and some writers, and it tends to deny the possibilities for a much more positive set of inte-grations and relationships between land-based activities and other rural goods and ser-vices. Lowe (2006: 42), for instance, argues in a 'think tank' publication that fails to treat agriculture at all, despite being entitled 'A new rural agenda', that

> a farm-centred view of rural development not only privileges a particular sector, but also perpetuates some essentially backward-looking myths about what drives rural economies. Rural development policies are pre-occupied with farm adjustment prob-lems. If the goal is to widen the base and vitality of the economies of rural areas, it is surely important that the crucial, consistent and largely non-agricultural drivers that are revitalising rural economies are supported.

2 It might be argued that some of these activities do in fact rely upon the growth of the consumption and post-productivist countryside (most clearly, forms of agri-tourism and some on-farm activities like farm shops). We have argued elsewhere that in order to contribute to a more multifunctional form of rural development these activities have to be transformed and meet at least three conditions: (i) add income and employment opportunities to the agricultural sector; (ii) contribute to the construction of a new mul-tifunctional agriculture that corresponds more effectively to the needs and expectations of the society at large; and (iii) imply a radical redefinition and reconfiguration of rural resources, to varying degrees both in and beyond the farm enterprise (Marsden and Sonnino 2005). In this sense, these rural development activities then attempt to inno-vate, add value and build upon a post-productionist trend by creating new synergies between agriculture and rural development, reintegrating agriculture both in terms of the rural economy and also addressing differentiated urban demands.

3 It is still quite extraordinary how official (Department of Environment, Food and Rural Affairs) reports on statistical trends on British agriculture still tend (i) to obscure spe-cific changes in farm structures and farm holdings, and (ii) by regular redefinition of

statistical boundaries (such as the size thresholds for what actually constitutes 'a farm') make any effective and large-scale longitudinal analysis of change ineffective. Two examples suffice. First, a recent DEFRA report, *Agriculture in the United Kingdom* (2004–5), devotes only seven pages (out of 125) of tables to the size and structure of the industry and does not deal with regional or temporal variations, focusing almost completely upon commodity changes. Second, changing definitions over time (such as registering small livestock holdings following the 2001 foot-and-mouth outbreak) tends to obscure the long-running agricultural concentration.

References

Barham, E. (2003) 'Translating *terroir*: the global challenge of French AOC labelling', *Journal of Rural Studies*, vol. 19, 127–38.

Bathelt, H. (2006) 'Geographies of production: growth regimes in spatial perspective 3: towards a relational view of economic action and policy', *Progress in Human Geography*, vol. 30, 223–36.

Buttel, F. (2006) 'Sustaining the unsustainable: agro-food systems and environment in the modern world', in P. Cloke, T.K. Marsden and P. Mooney (eds) *Handbook of Rural Studies*, London: Sage, pp. 213–30.

Cochrane, W. (1979) *The Development of American Agriculture: A Historical Analysis*, Minneapolis: University of Minnesota Press.

Day, G. (2006) 'Local food cultures', RELU Conference Paper, Advantage West Midlands, Aston Business School, May 2006.

Department of Environment, Food and Rural Affairs (2004–5) *Agriculture in the United Kingdom*, London: The Stationery Office.

Department of Environment, Food and Rural Affairs (2006) *Diversification in Agriculture, National Statistics*, London: The Stationery Office.

Department of Environment, Transport and the Regions/Ministry of Agriculture, Fisheries and Food (2000) *Rural White Paper for England*, London: The Stationery Office.

Du Puis, E.M. and Goodman, D. (2005) 'Should we go home to eat? Towards a reflexive politics of localism', *Journal of Rural Studies*, vol. 21, 359–71.

Friedmann, H. (2006) 'From colonialism to green capitalism: social movements and the emergence of food regimes', in F. Buttel and P. McMichael (eds) *New Directions in the Sociology of Rural Development*, Research in Rural Sociology and Development series, vol. 11, Amsterdam: Elsevier, pp. 227–65.

Frouws, J. and Mol, A. (1999) 'Ecological modernisation theory and agricultural reform', in H. De Haan and N. Long (eds) *Images and Realities of Rural Life*, Assen, the Netherlands: Van Gorcum, pp. 269–86.

Guzman, E.S. and Martinez-Alier, J. (2006) 'New social movements and agroecology', in P. Cloke, T.K. Marsden and P. Mooney (eds) *Handbook of Rural Studies*, London: Sage, pp. 472–84.

Hatannaka, M., Bain, C. and Busch, L. (2006) 'Differentiated standardization, standardized differentiation: the complexity of the global agrifood system', in T.K. Marsden and J. Murdoch (eds) *Between the Local and the Global: Confronting Complexity in the Contemporary Agri-food Sector*, Rural Sociology and Development series, vol. 12, Amsterdam: Elsevier, pp. 39–69.

Hendrickson, M.K. and Heffernan, W.D. (2002) 'Opening spaces through relocalisation: locating potential resistance in the weaknesses of the global food system', *Sociologia Ruralis*, vol. 42, 347–69.

Higgins, V. and Lawrence, G. (eds) (2005) *Agricultural Governance: Globalization and the New Politics of Regulation*, London: Routledge.

House of Commons (2004) *Milk Pricing in the UK, Environment, Food and Rural Affairs Committee, Ninth Report, 2003–4 HC 335*, London: The Stationery Office.

Ilbery, B. and Watts, D. (2004) 'Agricultural restructuring the West Midlands region, 1950–2002', *Journal of the Royal Agricultural Society of England*, vol. 165, 1–9.

Knickel, H. and Renting, M. (2000) 'Methodological and conceptual issues in the study of multifunctionality and rural development', *Sociologia Ruralis*, vol. 40, 512–18.

Lobley, M. and Potter, C. (2004) 'Agricultural change and restructuring: recent evidence from a survey of agricultural households in England', *Journal of Rural Studies*, vol. 20, 499–510.

Lobley, M., Errington, A., McGeorge, A., Millard, N. and Potter, C. (2002) *Implications of Changes in the Structure of Agricultural Businesses*, research report prepared for Department of Environment, Food and Rural Affairs: Final Report, Department of Land Use and Rural Management, Seale-Hayne Campus, University of Plymouth.

Lowe, P. (2006) 'European agricultural and rural development policies for the 21st century', in J. Midgley (ed.) *A New Rural Agenda*, Newcastle: Institute for Public Policy Research, pp. 29–46.

Lowe, P., Clark, J., Seymour, S. and Ward, N. (1997) *Moralizing the Environment: The Environment, Countryside Change, Farming and Pollution*, London: UCL Press.

Marsden, T.K. and Murdoch, J. (eds) (2006) *Between the Local and the Global: Confronting Complexity in the Contemporary Agri-food Sector*, Research in Rural Sociology and Development series, vol. 12, Amsterdam: Elsevier.

Marsden, T.K. and Parrott, N. (2006) 'Reconnecting farming and the countryside', in D. O'Connor, H. Renting, M. Gorman and J. Kinsella (eds) *Driving Rural Development: Policy and Practice in Seven European Union Countries*, Assen, the Netherlands: Van Gorcum, pp. 201–33.

Marsden, T.K. and Smith, E. (2005) 'Ecological entrepreneurship: sustainable development in local communities through quality food production and local branding', *Geoforum*, vol. 36, 440–51.

Marsden, T.K and Sonnino, R. (2005) *Rural Development and the Project State: Denying Multifunctional Agriculture in the UK*, European Union MULTIAGRI project report (contract 505297).

Marsden, T.K., Munton, R.J.C., Whatmore, S. and Little, J. (1986) 'Towards a political economy of capitalist agriculture: a British perspective', *International Journal of Urban and Regional Research*, vol. 10, 498–521.

Marsden, T.K., Murdoch, J., Lowe, P., Munton, R.J.C. and Flynn, A. (1993) *Constructing the Countryside*, London: UCL Press.

Morgan, K., Marsden, T.K. and Murdoch, J. (2006a) *Worlds of Food: Power, Place and Provenance in the Food Chain*, Oxford: Oxford University Press.

Morgan, K. *et al.* (2006b) *Going Local: Regional Innovation Strategies and the New Agri-food Paradigm*, final report to the Economic and Social Research Council (RES-000-23-0056).

Munton, R. and Marsden, T.K. (1991) 'Dualism and diversity in family farming? Patterns of occupancy change in British agriculture', *Geoforum*, vol. 22, 105–17.

Murdoch, J., Lowe, P., Ward, N. and Marsden, T.K. (2003) *The Differentiated Countryside*, London: Routledge.

O'Conner, D., Renting, H., Gorman, M. and Kinsella, J. (eds) (2006) *Driving Rural Development: Policy and Practice in Seven EU Countries*, Assen, the Netherlands: van Gorcum.

Potter, C. (2004) 'Multifunctionality as an agricultural and rural policy concepts', in F. Brouwer (ed.) *Sustaining Agriculture and the Rural Environment: Governance, Policy and Multifunctionality*, Cheltenham: Edward Elgar, pp. 15–35.

Potter, C. and Burney, J. (2002) 'Multifunctionality in the World Trade Organization: legitimate non-trade concern or disguised protectionism?', *Journal of Rural Studies*, vol. 18, 35–47.

Renting, H., Marsden, T.K. and Banks, J. (2004) 'Understanding alternative food networks: exploring the role of short supply chains in rural development', *Environment and Planning A*, vol. 35, 393–411.

Rogaly, B. (2006) 'Intensification of work-place regimes in UK agriculture: the role of migrant workers', Sussex Migration Working Paper 36, July 2006.

Sage, C. (2003) 'Social embeddedness and relations of regard: alternative "good food" networks in south-west Ireland', *Journal of Regional Studies*, vol. 19, 47–60.

Smith, E., Marsden, T.K., Flynn, A. and Percival, A. (2004) 'Regulating food risks: rebuilding confidence in Europe's food?', *Environment and Planning C*, vol. 22, 543–67.

Sonnino, R. (in press) 'Embeddedness in action: saffron and the making of the local in Southern Tuscany', *Agriculture and Human Values*.

Sonnino, R. and Marsden, T.K. (2006) 'Alternative food networks in the south west of England: towards a new agrarian eco-economy?', in T.K. Marsden and J. Murdoch (eds) *Between the Local and the Global: Confronting Complexity in the Contemporary Agri-food Sector*, Research in Rural Sociology and Development series, vol. 12, Amsterdam: Elsevier, pp. 299–323.

Stuiver, M. (2006) 'Highlighting the retro side of innovation and its potential for regime change in agriculture', in T.K. Marsden and J. Murdoch (eds) *Between the Local and the Global: Confronting Complexity in the Contemporary Agri-food Sector*, Research in Rural Sociology and Development series, vol. 12, Amsterdam: Elsevier, pp. 299–323.

Sylvander, B. *et al.* (2005) 'The strategic turn of organic farming in Europe: a resource-based approach for organic farming initiatives', conference paper: Organic Agriculture: Researching Sustainable Food Systems, Australia, 21–23 September, 2005.

UK Government (2002) *Policy Commission on the Future of Farming and Food: A Sustainable Future*, London: Cabinet Office.

Van der Grijp, N., Marsden, T.K. and Barbarosa-Cavakanti, J.S. (2005) 'European retailers as agents of change towards sustainability: the case of fruit production in Brazil', *Environmental Sciences*, vol. 2, 31–46.

Van der Ploeg, J.D. (2000) 'Revitalizing agriculture: farming economically as a starting ground for rural development', *Sociologia Ruralis*, vol. 40, 497–511.

Van der Ploeg, J.D. (2003) *The Virtual Farmer*, Assen, the Netherlands: van Gorcum.

Walford, N.(2003a) 'A past and a future for diversification on farms? Some evidence from large-scale, commercial farms in south-east England', *Geografisker Annaler B*, vol. 85, 51–62.

Walford, N. (2003b) 'Productivism is allegedly dead, long live productivism: evidence of continued productivist attitudes and decision-making in south-east England', *Journal of Rural Studies*, vol. 19, 491–501.

Walford, N. (2005) 'Agricultural restructuring during the closing decades of the twentieth century: evidence of farm size in south-east England', *Geography*, vol. 90, 238–49.

Ward, N. (1993) 'The agricultural treadmill and the rural environment in the post-productivist era', *Sociologia Ruralis*, vol. 33, 348–64.

Ward, N. (2006) 'Rural development and the economies of rural areas', in J. Midgley (ed.) *A New Rural Agenda*, Newcastle: Institute of Public Policy Research, pp. 46–67.

Watts, D.C., Ilbery, B. and Maye, D. (2005) 'Making reconnections in agro-food geography: alternative systems of food provision', *Progress in Human Geography*, vol. 29, 22–40.

Wilson, G. (2001) 'From productionism to post-productionism ... and back again? Exploring the (un)changed natural and mental landscapes of European agriculture', *Transactions of the Institute of British Geographers*, vol. 26, 23–52.

Winter, D.M. (2003) 'Embeddedness, the new food economy, and defensive localism', *Journal of Rural Studies*, vol. 19, 23–32.

Wrigley, N. (2000) 'The globalization of retail capital: themes for economy geography', in G.L. Clark, M.P. Feldman and M.S. Gertler (eds) *The Oxford Handbook of Economic Geography*, Oxford: Oxford University Press, pp. 292–313.

4 Revisiting landownership and property rights

Michael Winter

Introduction

The study of landownership and property rights was very much to the fore in the early years of the 'new' rural studies that began to emerge in sociology, geography and planning departments in the late 1970s, and Richard Munton was one of those who helped to galvanise and carry forward this interest. He was not alone in championing studies of landownership, with the study of rural landownership issues in Suffolk by Newby and colleagues still a seminal work for students in this area, and a small but significant number of other researchers entering this territory (Newby *et al.* 1978; Marsden 1984, 1986; Whatmore 1986).

This chapter starts by examining some of Munton's concerns on this topic and how they helped shape his early career. So central were land occupancy issues to that period of Munton's work that it is an obvious step then to look back to the origins of the land issue in British public life. There then follows a brief examination of how the land question has been dealt with in academia in the period following that initial explosion of interest in the late 1970s and early 1980s. Despite the existence of a small number of useful studies, remarkably little academic attention has, in fact, been given to landownership and property rights in Britain since that period. It is possible to provide a brief analysis of some of the key trends in British landownership and occupancy over the past 20 years, but the implications for environmental management and rural development remain only sketchily understood. The chapter concludes with some reflections on the state of rural studies in this context.

Landownership in rural studies

Landownership and land tenure form one of the enduring themes of Richard Munton's academic career, and his early work on these topics did much to revive wider social science interest in a subject that had in large part become the preserve of technical specialists in agricultural economics and land management by the 1960s and 1970s. It is a theme that encapsulates better than any other *both* his intellectual development and transition *and* his ability to combine a high level of academic sophistication and integrity with policy engagement. His earliest published research papers lie in this area, and for those familiar only with Munton's later work within a broadly political economy genre of geography, it will perhaps come as a surprise to see work not in behavioural geography – the precursor to the 'political economy turn' for several rural geographers in the 1970s – but an approach rooted in neo-classical economics. The content of these papers would seem to have pointed towards a career in agricultural economics rather than in mainstream geography at this stage. Indeed, one was published in an agricultural economics journal, a paper in which Munton (1975) dissects the agricultural land market in the early 1970s through an analysis of auctioneers' property transactions. Replete with 21 numeric tables, this is a painstaking paper that seeks to unravel some of the complexities of the land market, particularly the lack of fit between recorded prices for land and agricultural value, with the former being higher than the latter. Munton cites two reasons for this discrepancy: first, the demand for hobby farms and second homes; and second, the existence of 'hope' value resulting from the prospect of planning permission for non-agricultural activities (a theme also made explicit in the second of these early papers; see Munton 1976). Thus, although there is perhaps little in these papers to foreshadow his subsequent methodological and conceptual interests, these two substantive issues are certainly suggestive of future interests in the planning system and in the relation between landownership and economic, social and subsequently *sustainable* development. These are foundations that inform Munton's seminal work on the green belt and his rather neglected but rigorously researched monograph on property development and landownership (Munton 1983, 1984; Goodchild and Munton 1985).

However, there are very few signs in the work of the 1970s and early 1980s of his subsequent more Marxian and sociological forays into land occupancy issues (Munton and Marsden 1991a, b). A telling example of this is the treatment, or rather lack of treatment, of the Marxist account of property and land by Massey and Catalano (1978). Published as early as 1978, this work receives only passing mention by Goodchild and Munton (1985) and, on each occasion, as an empirical referent rather than for theoretical or conceptual insight.

Richard Munton's undoubted grasp of the empirical intricacies of the British land market, as evident in these early publications, propelled him into the kind of policy engagement that usually comes to academics at a later career stage, if at all. In September 1977, he became one of two specialist advisers to the government's Northfield Committee, established to inquire into the acquisition and occupancy of agricultural land. The committee's work continued until May 1979, so this was a sustained period of policy engagement. Munton reflected subsequently on the experience in a published paper that set out to explain why committees of inquiry are an 'unlikely means of promoting policy change on contentious policy issues', clearly an indication of some frustration with a committee with such a wide range of opinion that many of the recommendations were tentative and two of the members submitted a minority report (Munton 1984: 167; Northfield 1979).

The chief focus of the Northfield Committee on uncovering the extent of the incursions of financial institutions into the land market was recognised by many, even at the time, as irrelevant to the main issues facing farmers, landowners and would-be farmers. The Northfield Report was a response to the concerns of the political left in the late 1970s not so much with land per se but more with the behaviour of financial capital. It was this that in part prompted Massey and Catalano's Marxist analysis, as well as more polemical calls on the left for the nationalisation of 'the City'.

The land question in British politics: a brief retrospect

Landownership and the 'landed interest' provided a constant backcloth to nineteenth-century British politics and was an issue by no means confined to the political left. It was a Conservative, Disraeli, who in 1880 declared that, alongside the governance of Ireland, the other main political concern of the day was 'the principles upon which the landed property of this country should continue to be established'. The land-reform movement that emerged in the last two decades of the century attracted interest and commitment from a spectrum of political opinion. Its immediate and direct results were few, and nor did taxation and local government reforms in the twentieth century produce the kind of wholesale changes in the control of land that had been advocated by John Stuart Mill, Henry George and others (Cannadine 1990). Lloyd George's ill-fated 1910 land tax was repealed by its originator in 1922, and Snowden's attempt to reinstate land taxation in 1931 lasted only until 1934. By the 1930s and 1940s, calls for land nationalisation had become very much an accepted part of the debate upon agricultural issues in Britain, but with little immediate prospect of success (Bateman 1989). The opportunity to introduce a limited programme of nationalisation was open to the Labour government elected in 1945, but instead it opted for security of tenure (Smith 1989). Land nationalisation has never again been seriously on the political agenda.

It is worth reflecting on just why the issues of land reform and nationalisation came to prominence. One thing *is* clear, and that is that for many commentators and campaigners, the issue had little to do with socialism. Some of the more radical and Utopian critics of landownership had more in common with anarchist ideologues such as Kropotkin (1898), and their sympathies were for petty peasant methods of farming rather than with state socialism. Thus, the 'back to the land' movement of the close of the century primarily involved a rejection of industrial and bureaucratic society per se, whether of capitalist or socialist hue, in favour of simple communal living. Of greater significance were the critics whose capitalist, indeed industrial, credentials were impeccable. Some, notably free-trade liberals such as John Bright and Joseph Chamberlain, attacked the distribution of landownership but not the underlying principles of the landlord–tenant system. Others, such as John Stuart Mill, advocated peasant proprietorship on the grounds of justice and equity and a strong belief in the peasantry as a counterbalance to the industrial masses (Dewey 1974). On land nationalisation, some, such as the one-time Minister of Agriculture Lord Addison, and the Oxford agricultural economist C.S. Orwin, clearly wished to see land nationalisation in order to further advance capitalist agriculture, rather than as a point of socialist principle. In the light of these mixed motives it is no surprise to find that Karl Marx (1881) was highly critical of those who advocated land nationalisation, condemning them for wishing to 'save capitalist rule and actually re-establish it'.

The point at issue throughout this period, sometimes explicitly but often implicitly, was the need to maintain and establish the conditions for a prosperous agriculture in the context of an industrial and commercial nation and in the light of a declining landowning class. The period from 1875 to 1939 was dominated by agricultural depression (with the exception of the war years) and by political debate regarding the role of the landowning class, with a growing assumption that the class could not survive. It is this latter point that is often forgotten in today's debate on land tenure, which ignores the context of the complex social, economic and political processes bringing about a decline but not the elimination of the British aristocracy (Beckett 1986; Cannadine 1990; Thompson 1963). In the light of today's pattern of mixed tenure and the growth of owner-occupation it is difficult to imagine the consternation that accompanied the expected decline of the landlord–tenant system at the outset of the twentieth century.

These broad-brush comments about big issues such as nationalisation and landlordism should not, however, detract attention from the emergence of a meso-politics of tenancy itself, and indeed it is this incrementalism in the politics of tenure that came to dominate. Thus, the issue that emerged in the closing decades of the nineteenth century as the central bone of contention was not security of tenure but tenant right, defined as 'the just claim of a farmer to compensation for any unexhausted improvements made by him and remaining on his holding at the end of his tenancy' (McQuiston 1973: 95). The Landlord and Tenant Act 1851 gave rights to tenants in certain circumstances, upon termination of a tenancy, to remove buildings they had constructed. But further legislation on tenant right in 1875 was ineffectual, as it rested upon the landowners' voluntary compliance. However, the Agricultural Holdings Act 1883 provided a statutory framework for compensation and extended to one year the period of notice to quit, thus opening the way to further legislation to protect tenants. After 1883 the whole issue of farm tenurial arrangements shifted in focus from social relationships on particular estates to national statutory provision. Further provisions for protecting the rights of tenants to compensation and for guaranteeing freedom of cropping came in 1890, 1900, 1906, 1908, 1913, 1914, 1919, 1920, 1922 and in the Agricultural Holdings Act 1923, which Densham (1989) refers to as the 'Magna Carta' of English agriculture. The 1923 Act consolidated the various pieces of legislation and laid down a full code for the rights of tenants to compensation. A fuller measure of security of tenure, however, did not come until the Agriculture Act 1947 and the Agricultural Holdings Act 1948, which provided full lifetime security of tenure even to existing short-term tenancies of two years or more. Subsequently, in 1976, Labour's Agriculture (Miscellaneous Provisions) Act extended security to two successions. The Agricultural Holdings Act 1984 provided the *opportunity* for fresh tenancies to be created for one life only, but did not rescind the rights of tenants granted under the 1976 Act (Nix *et al.* 1987; Rodgers 1985; Troup 1984; Marsden 1986).

Land tenure since the late 1980s

As indicated earlier, despite the impetus established by Newby, Munton and others, academic interest in land-tenure issues declined sharply in the late 1980s. But this statement needs some qualification. Inevitably, some of the earlier interest was maintained in later publications. For example, the issue was not entirely discarded in work involving Munton himself, particularly work emerging from the Countryside Change programme of research based at University College London (Marsden *et al.* 1993; Munton 1995; Whatmore *et al.* 1990). Moreover, the event of occupancy change, as opposed to tenure per se, became an important focus in research seeking to explain the pattern of environmental change in agriculture (Munton and Marsden 1991b; Potter and Lobley 1996). Notwithstanding these important exceptions, and a few others referred to in the remainder of this section, the research effort on rural property ownership and occupancy issues has both contracted and fragmented. And yet the substantive issues remain. Indeed, since the 1980s there have been major changes in legislation on agricultural holdings as a result of the introduction of farm business tenancies in England and Wales; successive policy reforms and associated changes in agricultural and land markets leading to a further growth of share and contract farming; a major land-reform process in Scotland; and, more recently, owner's rights to exclusive use of land being eroded by the extension of public access rights under the Countryside and Rights of Way Act 2000. This 2000 Act and the Scottish land-reform story are outside the scope of this chapter (but see Parker and Ravenscroft 2001, and Stockdale *et al.* 1996).

One of the few issues identified by the Northfield Committee to find any real echoes in the new Conservative free-market political orthodoxy that established itself so rapidly in the 1980s was that of the barrier to new entrants posed by the agricultural holdings legislation. The Conservative rhetoric of the 1980s was built around freedom of enterprise. For some in the party, agricultural holdings legislation was anathema, an undue interference in the marketplace that, among other things, restricted the overdue flow of new entrepreneurial blood into this heavily subsidised sector. However, other than the CAP itself, agricultural reform was hardly a high issue on the political agenda during an era of high unemployment and the poll tax. Indeed, a number of other vestiges of agrarian corporatism, such as the Agricultural Wages Board, survived the Tory years intact. Moreover, the Conservatives were particularly loath to tackle a problem without the prior agreement of the main interested parties, especially when two potentially opposed parties were both 'natural' party supporters (in this case the farmer and landowner lobby organisations).

It took some years for the stalemate to be broken, and this occurred only when incontrovertible evidence was provided of the extent to which the law was becoming out of step with land-tenure realities. The Royal Institution of Chartered Surveyors (RICS), whose members stood to gain from either a tightening or a freeing up of the legislation, instigated an academic inquiry into the state of agricultural tenure in England and Wales. This research, which I led, uncovered a much higher rate of unconventional tenure than anyone had predicted, representing a rare example, in my experience, of applied research findings giving genuine surprise to the client community (Winter *et al.* 1990). The broad outlines of the findings are given in Table 4.1.

The data were drawn from postal survey responses from 1,790 holdings covering 2.8 per cent of the total agricultural area. In terms of the overall balance between tenanted and owned land, the findings were broadly consistent with the results of the annual agricultural census. In the June 1988 Census, 37 per cent of the agricultural area of England and Wales was tenanted, compared to 41 per cent in the RICS survey. As the RICS survey included grass-keep land in the tenanted category, amounting to nearly 3 per cent of the total land, the two figures were in reality almost identical. On the one hand, this provided a high

Table 4.1 Total agricultural land area by tenure type (weighted data raised to national level), England and Wales

	Area (ha)	Percentage area
Summary		
Owner-occupied	6,959,057	58.7
Tenanted	4,891,291	41.3
Total	11,850,348	100.0
Detailed breakdown of 'tenanted' sector:		
Full agricultural tenancy with no share in ownership	3,204,484	27.0
Full agricultural tenancy with share in ownership	461,452	3.9
Total full agricultural tenancy	3,665,936	30.9
Formal unconventional		
MAFF-approved letting/licence	69,427	0.6
Gladstone v. Bower agreement	100,225	0.8
Partnership	269,668	2.3
Share farming	95,004	0.8
Total formal unconventional	534,324	4.5
Informal unconventional		
Gentleman's or informal agreement	209,324	1.8
Grass keep	304,016	2.6
Cropping licence	11,902	0.1
Sub-tenancy	24,743	0.2
Other	141,046	1.2
Total informal unconventional	691,031	5.9

degree of confidence in the statistical validity of the findings, but on the other hand, the extra detail uncovered in the RICS survey highlighted just how misleading the Ministry's crude data on tenancy could be. There were two main grounds for concern. First, much land on full agricultural tenancy is in fact held by tenants with a share in ownership (some 13 per cent of fully tenanted land), as shown in earlier research by Rose *et al.* (1977). Much of this land is held in family trusts and therefore might be treated as *de facto* owner-occupation. Second, the results showed that while conventional owner-occupation and full agricultural tenancies predominated, a significant minority of land was held under other arrangements, amounting to 10.4 per cent of all land in total.

The terms 'formal unconventional tenures' and 'informal unconventional tenures' were coined to distinguish between two basic groups of other tenurial arrangements. Excluding those cases where the 'tenant' had a share in ownership of land under full agricultural tenancies, the ratio of land under full agricultural tenancy to land under unconventional tenures was 2.6:1. For every 2.5 hectares of conventionally let land, the RICS survey showed there to be one hectare of unconventionally let land. This was a remarkable figure, suggesting a widespread need to resort to mechanisms of land occupancy outside the current provision of the holdings legislation. As one respondent to the postal survey rather graphically expressed it: 'The present law must have totally failed when for a solution gaps in the legislation are used rather than the legislation itself.' Most unconventional tenures were held by larger farmers and were combined with other types of tenure. Thus, very few unconventional tenancies exist as the sole form of occupancy for the farmer in question.

The report received wide publicity and was used by the RICS and other stakeholders in the debate that ensued on reforming the agricultural holdings legislation. The Country Landowners Association (CLA) was already committed to the continuation of the landlord–tenant system through enhanced possibilities for landlords to let land on less secure tenancies. A survey of its members in 1989 had shown 'a continuing interest in the landlord/tenant system' that was frustrated 'by a number of factors, amongst which agricultural holdings legislation is pre-eminent'. The Association favoured freedom of contract because 'young men would stand a higher chance of getting a farm because there would be more to tender for'. The CLA's position was in accordance with the Conservative government's commitment to market forces in other areas of the economy, and the Association appealed to the government to 'lift the dead hand of the current tenancy laws in respect of new contracts'.

Much therefore hinged on the National Farmers' Union (NFU), highly suspicious traditionally of attempts to circumvent the problems posed by security of tenure. In February 1990 Sir Simon Gourlay, the NFU's president (speaking at a conference jointly organised by the Royal Agricultural Society of England, the Agricultural Development and Advisory Service and the RICS) acknowledged that he had been surprised by initial findings of the RICS survey presented by me at a special meeting called by the RICS a month earlier, and a show of hands at the NFU AGM in January 1990 indicated a new willingness to look at fresh options. Research findings had clearly, on this occasion, influenced politics. But it was to be a slow process. In April 1990 the NFU issued a joint statement with the Tenant Farmers Association, the Farmers' Union of Wales and the Association of County Councils, which gave renewed emphasis to defending tenant rights. However, the statement made one absolutely crucial concession in calling for 'a modern tenancy system [that] must allow the flexibility to give tenants access to diversification'. This need to loosen up tenancy restrictions to allow tenants the opportunity to diversify into non-agricultural enterprises on their holdings provided for a point of commonality between landowners and tenants. It required legislative change and it was the single most important factor in allowing negotiations to take shape over the following two years. The Agricultural Tenancies Act (1995) allowed for the introduction of much freer farm business tenancies, and the impact of this legislation has been studied in a number of projects (TRIG 2003; Whitehead *et al.* 2002). As a result of this research on farm business tenancies, their extent and nature are reasonably well understood. However, the implications for the older forms of unconventional tenure and for newer share/contract farming arrangements are far less clear. Moreover, it is generally accepted that the agricultural census has become less reliable in recent years, owing to a declining response rate, and therefore even the crude land-tenure data available from the census are not a particularly good guide to the real nature of tenure. It is hypothesised that changes are now occurring rapidly as a result of the introduction of the Single Farm Payment, particularly through new share and contract farming arrangements. Some of these issues are being explored by myself, Mark Overton and Liz Griffiths at Exeter in a project funded by the Economic and Social Research Council project on the history of share farming in England.

Conclusions: property in rural studies

This final section starts with a brief polemic directed against some of the impacts of the 'cultural turn' within rural studies. The remarkable lack of attention to rural property rights and land occupancy in rural studies since the early 1990s is in large part a consequence of the 'cultural turn' with its postmodern distrust of meta-narratives (such as political theories of property) and its disdain for applied policy-related research (such as work on landownership and land management). A fragmentation, and indeed sometimes denial, of rural studies has been the consequence, and at a time when the sustainability agenda (broadly defined) has thrust rural land use and economic activity centre stage, there is a profound irony in the fact that much in rural studies has retreated to the margins. A notable exception to this has been the work of Ravenscroft (1999), one of the few contemporary writers within rural studies to pick up on the property theme, and even give it a 'cultural' twist through an analysis of the changing nature of farm tenure as a progression from traditional feudalism via 'neo-feudalism' to a post-feudal and even postmodern position (Gibbard *et al.* 1999). But this work is very much an exception.

And yet the evidence that forms of occupancy are critical to sustainability outcomes – economic, environmental and social – remains overwhelming. It is axiomatic, therefore, that knowledge for sustainability requires knowledge of property, both empirical and conceptual. There are two main aspects to any consideration of property rights. First, it is important to understand the justifications given for property in political theory. Emel *et al.* (1992) identify five main sources of justification: the labour theory of property associated with Locke; arguments from liberty associated with Mill; arguments from utility associated with Hume and Bentham; first occupancy principles associated with Rousseau and Kant; and arguments related to moral character, such as the notion of good stewardship. Understanding how property rights work in practice requires an understanding of these different ideological justifications and how they may or may not be reflected in specific legal arrangements. Second, whatever the nature of justificatory ideologies, property rights take practical form in terms of both law *and* social and economic behaviour. Property rights can only be truly understood if these elements are held in some tension in theoretical endeavour. Political theories of property need to be considered alongside political practices.

In other words, contrasting tenurial arrangements have different implications for the social and economic relationships between different groups of people. In recognising this, Whatmore *et al.* (1990) have classified agricultural property rights on a continuum from simple owner-occupation to contract farming, with the various shades of secure and insecure tenancy in between. They identify three main rights to land – ownership, occupation and use – and the distribution of these rights is reflected in the contrasting tenurial arrangements that have evolved. Under simple owner-occupation, rights of ownership, occupation and use are combined under the heading of agricultural capital. The farmer has control in each area. But as one moves along the spectrum, landed capital assumes responsibility for certain rights. Under an insecure tenancy, for example, landed capital has owner and occupier rights, with agricultural capital retaining user rights. Under contract farming, landed capital retains all land property rights, with the contract farmer responsible only for non-land inputs, such as labour and capital. This is a useful classification of the range of options and their implications for possible social relations and the distribution of power and responsibility. However, as the RICS survey work demonstrated, such a schema can be misleading, as the precise content of a particular arrangement is all-important in determining its nature. There is a danger that methodological problems analogous to those surrounding *de jure* versus *de facto* ownership will re-emerge unless the precise contents of tenancy arrangements are analysed very carefully. For example, contract farming may *de jure* vest all the property rights outside the hands of the farmer, but *de facto* the terms of the agreement may place considerable rights with the farmer. If one looks back to agricultural tenure prior to the arrival of secure tenancies, it is quite clear that under freedom of contract the relative power, not to mention income, of landlord and tenant fluctuated according to the profitability of agriculture, and tenurial forms evolved to deal with this issue (Offer 1991). During periods of depression it was often the tenant who could demand favourable rent reductions and improved conditions of leases.

Coming at the issue from the perspective of modern political economy, Whatmore *et al.* (1990) pay little attention to other, more traditional attempts to deal with the tenure question. Putting aside the insightful work of Denman (1957, 1958; Denman and Prodano 1972; Denman and Stewart 1959), it was as early as 1941 that Kelso delivered a plea for a property-rights approach to tenure within agricultural economics (Kelso 1941; see also Higgs 1972). Currie (1981), in developing Kelso's framework, identifies a distinction between ownership and operating structures. This leads to a classification based on differentiation according to ownership of land, ownership of labour and the provision of entrepreneurship. Ironically, Currie, a neo-classical economist, produced a classification that arguably has a more sociological flavour than that of Whatmore *et al.* (1990). This is because Currie is particularly interested in the implications of alternative tenurial arrangements for farm decision-making. He contends that implications for decisions and management cannot be simply read off from tenure without, in addition, considering the role of labour and capital. Thus, pure owner-occupation *may* produce highly efficient and productive farming on capitalist principles based on the use of hired labourers *or* peasant proprietorship, in which decisions on the allocation of family labour will depend upon factors other than the maximisation or even optimisation of profit. Nowadays, it is essential to add a third category to that of capitalist producer and peasant farmer: that of residential proprietor, whose decisions on agricultural land use are likely to be based on a view of property as an item of consumption rather than production, a positional good. This has been noted by Offer (1991) for the nineteenth century too.

It is my thesis that landownership and property rights remain as an important a line of inquiry now as when Richard Munton embarked on work on this topic in the 1970s. There are both empirical and conceptual gaps in our understanding that need research attention. If the challenges around climate change and sustainable development are to be met, there is a need to revisit these issues, and to do so not through the relativist lenses of postmodernism but through a re-engagement with the social science thinking that emerged to tackle property-ownership issues in the 1970s and 1980s but has been so greatly neglected in recent years.

References

Bateman, D.I. (1989) 'Heroes for present purposes? A look at the changing idea of communal land ownership in Britain', *Journal of Agricultural Economics*, vol. 40, 269–89.

Beckett, J.V. (1986) *The Aristocracy in England 1660–1914*, Oxford: Basil Blackwell.

Cannadine, D. (1990) *The Decline and Fall of the British Aristocracy*, New Haven, CT: Yale University Press.

Currie, J.M. (1981) *The Economic Theory of Agricultural Land Tenure*, Cambridge: Cambridge University Press.

Denman, D.R. (1957) *Estate Capital*, London: Allen & Unwin.

Denman, D.R. (1958) *Origins of Ownership*, London: Allen & Unwin.

Denman, D.R. and Prodano, S. (1972) *Land Use: An Introduction to Proprietary Land Use Analysis*, London: Allen & Unwin.

Denman, D.R. and Stewart, V.F. (1959) *Farm Rents*, London: Allen & Unwin.

Densham, H.A.C. (1989) *Scammell and Densham's Law of Agricultural Holdings*, London: Butterworth.

Dewey, C.J. (1974) 'The rehabilitation of the peasant proprietor in nineteenth-century economic thought', *History of Political Economy*, vol. 6, 17–47.

Emel, J., Roberts, R. and Sauri, D. (1992) 'Ideology, property, and groundwater resources', *Political Geography*, vol. 11, 37–54.

Gibbard, R., Ravenscroft, N. and Reeves, J. (1999) 'The popular culture of agricultural law reform', *Journal of Rural Studies*, vol. 15, 269–78.

Goodchild, R. and Munton, R.J.C. (1985) *Development and the Landowner: An Analysis of the British Experience*, London: George Allen & Unwin.

Higgs, R. (1972) 'Property rights and resource allocation under alternative land tenure forms a comment', *Oxford Economic Papers*, vol. 24, 428–31.

Kelso, M.M. (1941) 'Needed research in farm tenancy', *Journal of Farm Economics*, vol. 23, 291–304.

Kropotkin, P. (1898) *Fields, Factories and Workshops*, London: Thomas Nelson.

McQuiston, J.R. (1973) 'Tenant right: farmer against landlord in Victorian England 1847–1883', *Agricultural History*, vol. 47, 95–113.

Marsden, T.K. (1984) 'Land ownership and farm organisation in capitalist agriculture', in T. Bradley and P. Lowe (eds) *Locality and Rurality: Economy and Society in Rural Regions*, Norwich: GeoBooks, pp. 129–45.

Marsden, T.K. (1986) 'Property–state relations in the 1980s: an examination of land-lord–tenant legislation in British agriculture', in G. Cox, P. Lowe and M. Winter (eds) *Agriculture: People and Policies*, London: Allen & Unwin, pp. 126–45.

Marsden, T.K., Murdoch, J., Lowe, P., Munton, R.J.C. and Flynn, A. (1993) *Constructing the Countryside*, London: UCL Press.

Marx, K. (1881) 'Letter to Friedrich Adoph Serge', 20 June, http://www.marxists.org/archive/marx/works/1881/letters.

Massey, D. and Catalano, A. (1978) *Capital and Land*, London: Edward Arnold.

Munton, R.J.C. (1975) 'The state of the agricultural land market 1971–1973: a survey of auctioneers' property transactions', *Oxford Agrarian Studies*, vol. 4, 111–30.

Munton, R.J.C. (1976) 'An analysis of price trends in the agricultural land market of England and Wales', *Tijdschrift voor Economische en Sociale Geografie*, vol. 67, 202–12.

Munton, R.J.C. (1983) *London's Green Belt: Containment in Practice*, George Allen & Unwin.

Munton, R.J.C. (1984) 'The politics of landownership: institutional investors and the Northfield Inquiry', in T. Bradley and P. Lowe (eds) *Locality and Rurality: Economy and Society in Rural Regions*, Norwich: GeoBooks, pp. 167–78.

Munton, R.J.C. (1995) 'Regulating rural change: property rights, economy and environment: a case study from Cumbria, U.K.', *Journal of Rural Studies*, vol. 11, 269–84.

Munton, R.J.C. and Marsden, T.K. (1991a) 'Dualism or diversity in family farming? Patterns of occupancy change in British agriculture', *Geoforum*, vol. 22, 105–17.

Munton, R.J.C. and Marsden, T.K. (1991b) 'Occupancy change and the farmed landscape: an analysis of farm-level trends, 1970–85', *Environment and Planning A*, vol. 23, 499–510.

Newby, H., Bell, C., Rose, D. and Saunders, P. (1978) *Property Paternalism and Power*, London: Hutchinson.

Nix, J., Hill, P. and Williams, N. (1987) *Land and Estate Management*, Chichester: Packard.

Northfield, Lord (1979) *Report of the Committee of Inquiry into the Acquisition and Occupancy of Agricultural Land*, Cmnd. 7599, London: Her Majesty's Stationery Office.

Offer, A. (1991) 'Farm tenure and land values in England, c.1750–1950', *Economic History Review*, vol. 44, 1–20.

Parker, G. and Ravenscroft, N. (2001) 'Land, rights and the gift: the Countryside and Rights of Way Act 2000 and the negotiation of citizenship', *Sociologia Ruralis*, vol. 41, 381–98.

Potter, C. and Lobley, M. (1996) 'Unbroken threads? Succession and its effects on family farms in Britain', *Sociologia Ruralis*, vol. 36, 286–306.

Ravenscroft, N. (1999) '"Post-feudalism" and the changing structure of agricultural leasing', *Land Use Policy*, vol. 16, 247–57.

Rodgers, C.P. (1985) *Agricultural Tenancies: Law and Practice*, London: Butterworth.

Rose, D., Newby, H., Saunders, P. and Bell, C. (1977) 'Land tenure and official statistics: a research note', *Journal of Agricultural Economics*, vol. 28, 69–75.

Smith, M.J. (1989) 'Land nationalisation and the agricultural policy community', *Public Policy and Administration*, vol. 4, 9–21.

Stockdale, A., Lang, A.J. and Jackson, R.E. (1996) 'Changing land tenure patterns in Scotland: a time for reform', *Journal of Rural Studies*, vol. 12, 439–49.

Tenancy Reform Industry Group (TRIG) (2003) *Tenancy Reform Industry Group Final Report*, London: Department for Environment, Food and Rural Affairs.

Thompson, F.M.L. (1963) *English Landed Society in the Nineteenth Century*, London: Routledge.

Troup, D.A.G. (1984) *Agricultural Holdings Act, 1984: The Practitioner's Companion*, London: Surveyor's Publication.

Whatmore, S. (1986) 'Landownership relations and the development of modern British agriculture', in G. Cox, P. Lowe and M. Winter (eds) *Agriculture: People and Policies*, London: Allen & Unwin, pp. 105–25.

Whatmore, S., Munton, R.J.C. and Marsden, T.K. (1990) 'The rural restructuring process: emerging divisions of agricultural property rights', *Regional Studies*, vol. 24, 235–45.

Whitehead, I., Errington, A., Millard, N. and Felton, T. (2002) *An Economic Evaluation of the Agricultural Tenancies Act 1995*, University of Plymouth Report to Department for Environment, Food and Rural Affairs.

Winter, M., Richardson, C., Short, C. and Watkins, C. (1990) *Agricultural Land Tenure in England and Wales*, London: Royal Institution of Chartered Surveyors.

5   Between earth and life

Refiguring property through bioresources

Sarah Whatmore

> On the horizon are a whole new set of claims to proprietorship ... [that] arise out of the very perception of hybrids, out of mixes of techniques and persons, out of combinations of the human and non-human, out of the interdigitation of cultural practices.
>
> (Strathern 1999: 122)

Introduction

It is not an exaggeration to say that it was a shared interest in the question of property that led to Richard Munton and me crossing paths. Lord Northfield's Committee of Inquiry into the Acquisition and Occupancy of Agricultural Land published its landmark report just months before I began my career as an undergraduate Geography student at UCL (Northfield Committee 1979). Richard's reputation as an expert on agricultural land market processes and trends (e.g. Munton 1975) had led to his appointment as a special adviser to the Northfield Committee, advising on 'the role of financial institutions in the farmland market'. But I did not become aware of this until later, when I had moved on to study for an M.Phil. in Planning at the Bartlett School, University College London, and, in the throes of an intellectual fascination with Marxist rent theory, was undertaking research for my Master's thesis on the role of financial institutions in the ownership of agricultural land. I cannot now recall reliably whether or not Richard and I met to discuss our common enthusiasm for institutional landownership at this point. I think that we did, not least because I attribute to him a number of key contacts in the London headquarters of agricultural land agents such as Savills and Jones Lang Wootton. I cannot imagine who else would have been so well connected in those circles. What is certain is that the first paper I ever had published records my debt to Richard, citing no fewer than three of his papers (Whatmore 1986; Munton 1977, 1984, 1985).

It is an association, and a debt, that has deepened over the years and gained particular intensity during the three years when I worked as a Research Associate with Richard, and Terry Marsden, on an ESRC project about the restructuring of English agriculture. This was an immensely exciting and productive period that stands out as such to this day. Among the many papers that we produced together, at least two were on property rights (see Munton *et al.* 1988; Whatmore *et al.* 1990). As that work expanded under the auspices of the Countryside Change Programme, and a new research team took shape, so this interest in landownership continued, interrogating its role in shaping agricultural landscapes and rural environments and exploring its place in the complex market and regulatory processes associated with land development and land-use change. Richard published on the contestation of mining developments in Cumbria, employing the notion of 'real' regulation to signal the importance of attending to the 'nitty-gritty' conduct of the process of development control (Munton 1995). Terry, together with Jon Murdoch, wrote on the politics of land-use change in Buckinghamshire – a subject to which Jon remained committed right up to his untimely death (Murdoch and Marsden 1994; Murdoch 2006). It has been, and continues to be, a persistent vein in my own research concerns too, pursued through projects as diverse as the curious logics of environmental 'planning gain', native title and Aboriginal land rights in Australia, and on the extension of intellectual property rights to plant genetic resources (Whatmore and Boucher 1993; Whatmore 2002).

What I want to suggest and develop in this chapter is a way of taking forward, conceptually and substantively, these shared interests in property rights and their importance to questions of agricultural industrialization and environmental regulation with which many of us continue to grapple. Inevitably this means breaching the confines of a 'rural' framing of the research enterprise, but in so doing, as with travelling in another country, one opens the frame to a wider set of theoretical resources, political energies and material practices in ways that make rurality both more potent and more porous. The chapter focuses on tensions between the territorializing and deterritorializing tendencies through which property relations are forged. It also addresses the importance of understanding the materiality of property in ways that make sense of the continuities and discontinuities between 'real' or landed forms of property right and 'intangible' or intellectual forms of property right. Such an approach extends some of the core strengths of Richard Munton's research agenda, evidenced in both his published work and collaborative exchanges, into the life science era, a time in which earth resources are being equalled, if not supplanted, by bioresources as the key medium of agricultural and environmental change.

The matter of property

It is hardly original to observe that property is one of the primary currencies, if not the primary currency, through which conversations between law and geography have been conducted and continue to take place (Holder and Harrison 2002). This should come as no surprise given the shared complicity of these two disciplines in the 'cartographies' of governance, commerce and science, and in the recalibration of a litany of imagined 'exterior' and/or 'prior' space-times of modernity – whether those be associated with the age of empire or with the rubric of global environmental management today. The importance of critical variants of such conversations has been to refuse to take either the letter of the law or the boundaries it inscribes as self-evident configurations of justice or space, insisting instead on their historical and cultural situatedness (see Blomley 1994).

The well-rehearsed confusion in everyday understandings of modern Western forms of private property is, in the vernacular of estate agents, to mistake property for its objects ('the property for sale') while encouraging its subjects ('the property owners') to imagine themselves unqualified in their possession. In practice, as legal theorists and judiciaries have been at pains to remind us, the assorted bundle of beneficial rights that constitute property has rather less to do with defining 'the relationship between a person and his [*sic*] things, than with the relationship that arises between persons with respect to things' (Ackerman 1977: 26). This point has been well taken by those working at the interface between geography, anthropology and critical legal studies. It informs important analytical shifts beyond a concern with the 'effects' of property rights on the balance of powers between preconstituted social subjects, to the ways in which they (re)configure the very mode of social agency and relationality and its territorialization in distinctive institutional forms ranging from the borders of nation-states and the compass of corporate markets to the calculus of the individual and the domain of the self (see, for example, Verdery and Humphrey 2004).

However, it is my contention that rather less attention has been paid to the 'things', constituted as stable entities, exploitable resources and transactable goods, than to the 'persons' in this relationship, and to the ways in which private property rights fragment the potentialities of 'things' into discrete objects amenable to new forms of socio-technical alignment, metamorphosing themselves in the process. As Marilyn Strathern notes in her collection of essays *Property, Substance and Effect* (1999), the affordances and propensities of 'things' shape the performance of persons and property rights in particular times and places quite as much as they are shaped by them. Grand theories and conventional wisdoms of various kinds have tended to cast private property in heroic terms, as a 'geo-metric' mantle levelling the rude earth before the irresistible and comprehensive advance of Capitalism, and/or Civilization, and/or Modernity, in which land and property become synonymous through the metric of surface area, notably an isotropic plane finitely divisible into mutually exclusive estates (Blomley *et al.* 2001). But a host of other biophysical elements and processes, such as air or water, render this spatial codification of 'real' property as a grid-like surface both unimaginable and impracticable. In a similar vein, ongoing efforts to extend the purchase of intellectual property rights (IPR) from the inert materializations of human creativity (such as books) with which, historically, they have been mainly concerned to those of '*living* things' are fraught with difficulties associated with the promiscuous potencies of their mutable corporeality (Boyle 1996).

I want, here, to address this imbalance in work on property by outlining a more 'hybrid' mode of analysis that attends as closely to the 'things' as to the 'persons' mobilized in the spatial assemblages of private property relations. In particular, I want to signal the complicity of property laws in settling and unsettling taken-for-granted distinctions between the provinces of the social and the material, the human and the non-human, through their shifting determination of what and/or who constitutes persons and things (*persona* and *res*). Perhaps the most pressing impetus for this endeavour is the fact that geography, like law, is having to come to terms with the conceptual and empirical challenges wrought by the life sciences and the rapid proliferation of phenomena, from global warming to the multifarious progeny of biotechnologies, that defy the routine purification of the world into the categorical encampments of *either* nature *or* culture (see, for example, Delaney 2003). Rural geography, of course, is not exempt from these new imperatives. Indeed, among the most influential works in the wider discipline's efforts to address itself to this context are those whose empirical interests include the practices of agriculture, the evolution of land use and the conservation of wildlife. While the flourishing of science and technology studies (STS) over the past decade or so has accorded scientific knowledge practices and their durable incarnation in a range of socio-technologies a primary place in the assemblage of these binary coordinates, I want to suggest that legal practices, and their durable incarnation in the codification of a range of property (and other) rights, are just as significant in 'making the cut' or drawing the line that seeks to mark the social from the natural.

Perhaps because geographers inhabit this settlement more self-consciously than members of most other disciplines, they have readily explored the spatial implications of the conceptual vocabularies elaborated in STS that subject the 'nature–society' settlement to constant interrogation by way of such disruptive and potent figures as those of the 'hybrid' or the 'cyborg' (see, for example, Whatmore 2002; Murdoch 2006). Moreover, geography, anthropology and STS have been more inclined than most disciplines to experiment with methodologies that more effectively track the power of science through the effacement of its own practices, a process that holds lessons for studying legal practices. Such methods attend to the laborious business of getting up close and observing the humdrum business of law, and science, in the making; tracking the assemblage of particular interpretive communities; the reification of particular words and phrases; the sedimentation of precedents and protocols in technologies of documentation and record; and the rituals of professional training, hierarchies and divisions of labour (see, for example, Manderson 2005). Like technology, the role of property can be understood as a mode of socio-material ordering which attaches people differentially to networks of heterogeneous others (human and non-human, living and inert) whose capabilities are intricately interwoven in the performance of its partial and precarious hold on the world (Rose and Novas 2005). Moreover, science and law, technology and property, invariably work in tandem in the assemblage of such hybrid artefacts and spaces as, for example, the alignment of the knowledge practices and devices of genetic engineering and the legal practices and precedents of IPR associated with the emergence of bio-industrialization (see, for example, Pálsson and Rabinow 1999; Hayden 2003).

As 'the basis of modern law' (Fitzpatrick 1992: 82), property can be seen to play a crucial part in governing the shifting coordinates of two of the primary axes of the modern political constitution, those of the 'inside' and the 'outside' of political community, territorialized in the nation-state and configuring who/what counts as a political (and legal) subject, and of 'society' and 'nature', territorialized as distinct ontological domains and configuring what/who constitutes a legitimate object of proprietorial claims (Strathern 1999). As scientific and legal efforts to render life amenable to exploitation as 'biological resources' threaten to mesmerize us by their apparent novelty, it is timely to recall that their contested reworking of these binary coordinates mimics and disrupts inherited narratives and practices of 'enclosure' and 'the commons' that have figured as the parable of modern political ordering since the time of Locke in the seventeenth century. At its starkest, the global restaging of these ordering narratives and practices in the early twenty-first century pits two competing impulses against each other: the reconstitution of the subjects and spaces of political community and legal rights associated with the generic claims of 'humankind', and the disintegration of the established borders between human and other kinds in the fashioning of biological resources and new forms of commodity.

In the limited scope of this chapter I cannot develop the 'close scrutiny' of property law in practice that I advocate above (but see Verdery and Humphrey 2004 for a range of just such detailed case studies). Instead, I want to illustrate the arguments that I have outlined through glimpses of two case studies elaborated elsewhere and amplified by other examples in the contemporary geographical literature (Whatmore 2002). In illustrating these arguments I shall open the frame of British rural geography to lessons from the ongoing legacy of Britain's colonial past and from forums of global governance in which British interests are negotiated and invested. My first case study concerns the so-called *Mabo* ruling in the Australian High Court in 1992, which marked a significant shift in the legal fiction underpinning the colonial 'settlement' of the continent and admitted the validity of 'native title' into Australian common law. The second interrogates the International Undertaking on Plant Genetic Resources (IUPGR), established in 1983 by a conference resolution of the Food and Agriculture Organization (FAO) in an attempt to constitute and regulate global transactions in PGR as a 'heritage of humankind'. My emphasis in this account is to illustrate the contested labours of division through which legal distinctions between persons and things, subjects and objects of property, de- or reterritorialize the boundaries of nature and society in fraught and unstable ways.

Unsettling the 'state of nature'

> Land is not simply a speculative surface but a layered and multivalent
> space, itself caught up in the propertied politics of particular places.
>
> (Blomley 2002: 577)

On 3 June 1992, four plaintiffs from the island of Mer off the coast of Queensland won their ten-year battle in the Australian high court to overturn the arcane legal doctrine of *terra nullius* that had legitimated the colonization by 'settle-ment' of lands deemed 'uninhabited', and to retrieve their own claims on the land from under its shadow. More than a convenient fiction, *terra nullius* pro-mulgated the continent's literal and metaphorical 'emptying out' in European cartographies as the inconvenient people encountered by settlers were incorpo-rated into the fabric of the land they coveted as inhabitants of 'the state of nature'. The diverse indigenous peoples of Australia found themselves trans-fixed by their prior presence in the shifting cosmologies of colonization at the junction between history and biology as a relic and thoroughly distant human kind. The contemporary legal commentator Emeric de Vattel ([1760] 1916: 37) made explicit this transposition from the company of society to that of nature in arguing that 'those people ... who having fertile countries, disdain to cultivate the earth and chose rather to live by rapine, are wanting to themselves and deserve to be exterminated as savage and pernicious beasts'.

The 1992 *Mabo* ruling rejected the doctrine and its later 'evolutionary' vari-ants that had continued to obviate the land rights of Aboriginal Australians in intervening High Court judgments through the twentieth century. The majority judges found that Aboriginal land rights had not been extinguished by the assumption of Crown sovereignty, but, more than this, these antecedent rights represented a 'burden' on the Crown from the first under the provisions of English common law. For the first time in Anglo-Australian jurisprudence the *Mabo* judgment disentangled sovereignty and property and opened up the space between the Crown's radical title to the territory of the colony, a matter of public law, and its absolute beneficial title to the land, the jurisdiction of the common law.

In their written judgment, the majority sought to underline that it had been the 'actions of Governments' (of the Australian colonies and, more recently, of the Commonwealth, States and Territories) rather than common law principles that had substantially dispossessed Australia's aboriginal peoples (ALR, *Mabo* v. *Queensland*, 1992: 434 *passim*). The judgment, and subsequent Native Title legislation enacted by the Commonwealth Government of Australia in 1993, dis-turbed the settled parameters of the law of the land and the place of the 'native' in the space of the nation. First, in terms of the land, two of the majority judges began by arguing against the 'assumption that the ownership of land naturally breaks itself up into estates, conceived as creatures of inherent legal principle' (Justices Dean and Gauldron, ALR 1992: 441, para. G). In place of the geo-metric calculus which casts land as a surface area contained by lines on a map, the majority judgment evoked a more fluid proprietory landscape complicated by a variety of property rights falling short of freehold, the gold standard of private property, that configure different kinds of entitlement to various substan-tive qualities and affordances of the land (e.g. minerals, pasture, game and aquifers), often in relation to the same area or plot. This cautious admission of 'proprietary pluralism' cast 'native' title claims within the same universe of particular and provisional land rights as a host of 'settler' titles, from pastoral leases and mining licences to squatter grants and game permits, and provided an unwelcome reminder to some of Australia's most powerful landed interests in the mining and pastoral industries that their own claims were less unqualified than they had accustomed themselves to assume.

Second, in terms of Aboriginal Australians' status as proprietary subjects the majority judges variously queried the privileged status of cultivation in European political and legal theory as the only hallmark of proprietary interest capable of recognition at common law. Citing precedents from North American courts and the International Court of Justice, Justice Toohey argued:

> It is presence amounting to occupancy which is the foundation of the [native] title and which attracts protection, and it is that which much be proved to establish title.... Not the occupation of a particular society or way of life.
>
> (ALR 1992: 486, para. C)

This broader definition of occupancy required proof of a meaningful connection to land 'from the point of view of the members of the society [of claimants]' (ALR 1992: 486, para. C). It opened the way to acknowledging more varied and robust forms of native title in accordance with traditional uses, rights and obligations irrespective of whether or not they conformed to 'English or European modes or legal notions' (Justices Deane and Gualdron, ALR 1992: 441, para. C). On the narrowest interpretation, the *Mabo* judgment and Native Title Act 1993 acknowledged the tenacious continuity of Aboriginal attachments to the land in the interstices of European 'settlement', a tenacity now confined to a minority of indigenous Australians and to some of the remotest parts of the country. Their larger significance lies in their inscription of difference and multiplicity into the fabric of Australian law and governance against the grain of precedent and in the face of ongoing political opposition.

This multivalency of private property rights even in that seemingly most settled of resources – land – is echoed in the insightful paper by Nick Blomley (2002) on the politics of land development in downtown Vancouver. Using the archaic legal language of disputed property claims, he argues that the 'quieting of title' tells us 'that a property regime doesn't just happen ... but requires a continual doing' (ibid.: 557). In this he underlines the lesson of the *Mabo* ruling too, that despite assumptions to the contrary, 'law's land' is in practice anything but quiet, namely securely possessed through determinate title and inert and quiescent in the performance of property relations.

Cultivating the limits of possession

Resolution 8/83 at the conference of the UN Food and Agriculture Organization in Rome in 1983 established the first attempt in international law to regulate transactions in the genetic properties of plant life in the guise of the International Undertaking on Plant Genetic Resources (IUPGR) and the associated regulatory infrastructure of the Commission for Plant Genetic Resources (CPGR) (Food and Agricultural Organization 1983). Specifically, the IUPGR set out to institutionalize PGR as a 'heritage of humankind', taking its cue from earlier episodes in global environmental governance that had seen the oceans, Antarctica and outer space (re)constituted through international treaties as the 'common heritage of humankind'. First established by the Law of the Sea, this legal principle has become the hallmark of so-called global commons – resources/spaces invested with a virtuous environmental hue as the last preserve of an original nature 'outside' the reach of rival claims to sovereignty or property rights. The objective of the FAO's attempt to extend this principle to PGR was to 'ensure that PGR of economic and/or social interest, particularly for agriculture, will be explored, preserved, evaluated and made available for plant breeding and scientific purposes' (FAO 1983, Article 1: 50). The political contours of this regulatory initiative centre on the de- or reterritorialization of plant biodiversity. Crudely put, the geopolitics of PGR then (and, arguably, now) concerned conflicting efforts by alliances of nation-states represented in the FAO conference chamber to secure or restrict the redistribution of PGR and its potential benefits from indigenous concentrations of species and genetic diversity in tropical and semi-tropical habitats primarily located in so-called developing countries, through the institutions of corporate science and patent law most actively engaged in its technological and commercial appropriation and predominantly located in North America or Western Europe.

The definition of PGR set out in the International Undertaking makes plain that its business is to determine and regulate the point at which plant germ plasm ceases to be deemed a 'gift of nature' and becomes transformed into a social artefact attributable to particular social agents. The 'technical' niceties of the categorizations of plant materials that qualify as PGR make the difficulties of this exercise evident and emphasize that the distinctions codified in the Undertaking are just that, artefacts of this particular annunciation, and neither inhere in nor exhaust the restless morphogenesis of plant life.

> PGR means the reproductive or vegetative propagating material of any of the following categories of plant:
>
> i cultivated varieties (cultivars) in current use and newly developed varieties;
> ii obsolete cultivars;
> iii primitive cultivars (land races);
> iv wild and weed species, near relatives of cultivated varieties;
> v special genetic stocks (including elite and current breeders' lines and mutants).
>
> (FAO 1983, Article 2(a): 50)

Like efforts to protect new plant varieties and plant breeders' rights before it (under the Union pour la Protection des Obtentions Végétales (UPOV)), the assemblage of PGR as a 'thing' amenable to appropriation rests on an alliance of specific socio-technical and socio-legal competences. The extension of patent claims to PGR is crucially dependent on the fusion of human/plant energies becoming *incarnate* in the 'thing' itself, such as in a distinctive cell line or DNA sequence.

The FAO's efforts to prevent the realignment of energies and redistribution of benefits associated with the patenting of PGR centred on pushing this question of the incarnation of hybridity, or what Strathern (1999: 128) calls the 'bodying forth' of social energies in plant life, beyond the patent criteria (of inventiveness, usefulness and novelty) established in Anglo-American jurisprudence which effectively privilege the knowledge practices/artefacts of Western science. As the IUPGR suggested, the incarnation of human/plant hybrids does not stop, or begin, there. What about the entanglements of plant and human energies through centuries of peasant seed selection, breeding and sowing? Under the IUPGR provisions these too were to be legally recognized in the guise of 'farmers' rights'. What then of the biological transformations wrought by communities practising nomadic or foraging livelihood strategies? Following a campaign by indigenous peoples' groups (backed by anthropologists), the CPGR began to formulate legal instruments for the recognition 'indigenous rights' claims too (see Posey and Dutfield 1993).

As a precursor to the Convention on Biodiversity, the IUPGR represents an episode in the geopolitics of biodiversity management that exposed the partiality of Western modes of IPR as the boundary markers between the natural and the social, a common heritage and an exclusive claim. With its contradictory objectives, of conserving biodiversity and aiding its commercial exploitation, the Convention represents a very different balance of political alliances and legal instruments. Thus far, it has sought to reduce the commons status of PGR to the legally meaningless 'common interest' of humankind and to impose patenting (regulated under the TRIPS (Trade-Related Aspects of Intellectual Property Rights) provisions of the WTO) as the 'required' method of protecting plant varieties and so on in all signatory countries (see Parry 2004). But the legacy of the IUPGR continues to haunt this settlement as a constant reminder that its attempt to draw the line between the social and the natural is, like those before it, anything but settled.

To some extent, perhaps, plant genetic resources are still within our 'comfort zone' when we consider the implications of the application of IPR to biological materials. However, any residual complacency is likely to be punctured by work by Beth Greenhough (2006) on the privatization of Iceland's medical records. She traces the processes through which a private company, deCODE, acquired a licence from the Icelandic government to build and exploit this centralized health sector database. In this, such work is vital to attuning geographers' sustained concerns with the social and environmental consequences of private property rights to

> the pervasive register and technologies of biological engineering which permeate the boundary that has been taken to mark off 'human society' from

the rest of the 'natural world' and reconstitute 'us' as much as 'it' as techni-
cally equivalent subjects in the vast informatic menagerie of life science.

(Spencer and Whatmore 2001: 140)

Conclusions

I began by linking the project of taking the 'things' constituted through property
relations as seriously as the 'persons' to the political and intellectual challenges
facing geography (and law) by the post-genomic space-times we inhabit. The geo-
metric calculus of landed property that has come to affirm the proper place of
nature and society in our cognitive and administrative maps of the world is being
increasingly disconcerted by more topological imaginaries and practices associated with the deterritorialization of the bodily integrity of living organisms and the
reterritorialization of discrete components: cells, genes, proteins as bioinformatic
resources. The terms on which hybridity is being codified in legal protocols like
those governing rights in, and jurisdictions over, plant germ plasm heralds a de- or
reterritorialization of the vital associations between plants and people no less con-
sequential than that effected through the calculus of cultivation for the lands
enclosed by European colonialism. Perhaps what brings these consequences home
most forcefully is precisely that they transgress common-sense boundaries of the
self and of the human and make us think again about our vulnerabilities and allegiances in a profoundly fabricated world. As legal theorists and practitioners start
to promote 'the law' as a means of addressing public anxieties by subjecting
biotechnological artefacts and risks to 'independent' scrutiny, the project advanced
here signals a note of caution, that law is as implicated as science in the constitution of these artefacts and risks, and should itself be subject to critical scrutiny.

This forward projection of geography's long-standing investment in the
importance and consequences of property rights regimes for environmental management and change owes much to the sustained research of Richard Munton
and bears the hallmark of three research principles that I have come to appreciate over the years as central to his work. The first of these is the importance of a
strong empirical base to research on property rights. As Richard urged over a
decade ago, 'detailed empirical inquiry must be given a comparable status to the
derivation of theoretical categories in the comprehension of change; both must
be continually informed by the other' (Munton 1995: 271). The second of these
principles is the central theoretical tenet that resources are made, not found.
Whether concerned with the exploitation of land for agriculture, mining or
development that he studied or, by extension, the bioresources that hold our
attention today, Richard Munton's research legacy reminds us that resources are
always socio-material imbroglios constituted through the performance of property relations in particular times and places. Last, but not least, his work continues to be one the best exemplars of the value of subjecting one's research to
the sobering question 'what difference does it make?', such that the social and
environmental implications of our research efforts remain a measure of its analytical purchase.

References

Ackerman, B. (1977) *Private Property and the Constitution*, New Haven, CT: Yale University Press.

Australian Law Reports (1992) *'Mabo* v. *Queensland'*, *Australian Law Journal*, vol. 66, 408–99.

Blomley, N. (1994) *Law, Space and Geographies of Power*, New York: Guilford Press.

Blomley, N. (2002) 'Mud for the land', *Public Culture*, vol. 14, 557–82.

Blomley, N., Delaney, D. and Ford, R. (eds) (2001) *The Legal Geographies Reader*, Oxford: Blackwell.

Boyle, J. (1996) *Shamans, Software and Spleen: Law and the Construction of the Information Society*, Cambridge, MA: Harvard University Press.

De Vattel, E. ([1760] 1916) *Le Droit des gens ou principes de la loi naturelle*, Washington, DC: Carnegie Institution.

Delaney, D. (2003) *Law and Nature*, Cambridge: Cambridge University Press.

Fitzpatrick, P. (1992) *The Mythology of Modern Law*, London: Routledge.

Food and Agriculture Organization (1983) 'Proposal for the establishment of an international genebank and preparation of a draft international treaty for PGR', Seventh Session of Committee on Agriculture (21–30 March 1983) CAOG/83/10, Rome: FAO.

Greenhough, B. (2006) 'Mapping the networks of bioinformatics exchange', *Environment and Planning A*, vol. 38, 445–63.

Hayden, C. (2003) *When Nature Goes Public: The Making and Un-making of Bioprospecting in Mexico*, Princeton, NJ: Princeton University Press.

Holder, J. and Harrison, C. (eds) (2002) *Law and Geography*, Current Legal Issues 5, Oxford: Oxford University Press.

Manderson, D. (ed.) (2005) ' Legal spaces', *Law, Text, Culture*, vol. 9 (special issue).

Munton, R.J.C. (1975) 'The state of the agricultural land market in England 1971–1973: a survey of auctioneers' property transactions', *Oxford Agrarian* Studies, NS, vol. 4, 111–30.

Munton, R.J.C. (1977) 'Financial institutions: their ownership of agricultural land in Great Britain', *Area*, vol. 9, 29–37.

Munton, R.J.C. (1984) 'The politics of rural landownership: institutional investors and the Northfield Inquiry', in T. Bradley and P. Lowe (eds) *Locality and Rurality*, Norwich: GeoBooks, pp. 167–78.

Munton, R.J.C. (1985) 'Investment in British agriculture by the financial institutions', *Sociologia Ruralis*, vol. 25: 155–73.

Munton, R.J.C. (1995) 'Regulating rural change: property rights, economy and environment: a case study from Cumbria, UK', *Journal of Rural Studies*, vol. 11, 269–84.

Munton, R.J.C., Whatmore, S. and Marsden, T. (1988) 'Reconsidering urban-fringe agriculture: a longitudinal analysis of capital restructuring on farms in the metropolitan green belt', *Transactions of the Institute of British Geographers*, vol. 13, 324–36.

Murdoch, J. (2006) *Poststructuralist Geographies*, London: Sage.

Murdoch, J. and Marsden, T.K. (1994) *Reconstituting Rurality: Class, Community and Power in the Development Process*, London: UCL Press.

Northfield, Lord (1979) *The Committee of Inquiry into the Acquisition and Occupancy of Agricultural Land*, Cmnd. 7599, London: Her Majesty's Stationery Office.

Pálsson, G. and Rabinow, P. (1999) 'Iceland: the case of a national human genome project', *Anthropology Today*, vol. 15, 14–18.

Parry, B. (2004) *Trading the Genome*, New York: Columbia University Press.

Posey, D. and Dutfield, G. (1996) *Beyond Intellectual Property*, Ottawa: International Development Research Centre.

Rose, N. and Novas, C. (2005) 'Biological citizenship', in A. Ong and S. Collier (eds) *Global Assemblages: Technology, Politics and Ethics as Anthropological Problems*, Oxford: Blackwell, pp. 439–63.

Spencer, T. and Whatmore, S. (2001) 'Bio-geographies: putting new life back into the discipline', *Transactions of the Institute of British Geographers*, vol. 26, 139–41.

Strathern, M. (1999) *Property, Substance and Effect: Anthropological Essays on Persons and Things*, London: Athlone Press.

Verdery, K. and Humphrey, C. (eds) (2004) *Property in Question: Value Transformation in the Global Economy*, Oxford: Berg.

Whatmore, S. (1986) 'Landownership relations and the development of modern British agriculture', in G. Cox, P. Lowe and M. Winter (eds) *Agriculture: People and Policies*, London: Allen & Unwin, pp. 105–25.

Whatmore, S. (2002) *Hybrid Geographies: Natures Cultures Spaces*, London: Sage.

Whatmore, S. and Boucher, S. (1993) 'Bargaining with nature: the discourse and practice of environmental planning gain', *Transactions of the Institute of British Geographers*, vol. 17, 166–78.

Whatmore, S., Munton, R.J.C. and Marsden, T.K. (1990) 'The rural restructuring process: emerging divisions of agricultural property rights', *Regional Studies*, vol. 24, 235–45.

6 Rurality and creative nature–culture connections

Paul Cloke

Commodifying rural space

Back in 1993 I contributed to a book of essays in honour of Allan Patmore, suggesting that the countryside was being commodified in new ways through the production of new spaces of tourism and leisure, and that as a result rurality was increasingly being experienced and 'known' through inauthentic spectacle and even sign-inverting simulacra (Glyptis 1993). These by now unremarkable observations were set against a background of a broader realisation on the part of rural geographers that rural–urban differences were becoming increasingly blurred (Bell 2005; Cloke 2005). The all-pervasive processes of cultural urbanisation and extended suburbanisation over previous decades had served to spread urban values and cultures out across the rural, while the popularity of urban 'villages' and suburban natures in the (re)presentation of the city had to some extent brought aspects of the rural into the urban. The very 'rural' characteristics and values that attracted in-migrants and visitors to rural living and leisure were being infiltrated by the commodifying processes of in-migration, leisure and tourism. As a consequence, imagined geographies of rurality were becoming as significant as, if not more significant than, the material geographical spaces of rurality, and relationships between society, space and nature in the countryside were becoming dynamic and complex.

Mormont's (1990) classic study of these changes identified five trends in rural society and space:

- an increase in the mobility of people, resulting in an erosion of the autonomy of local communities;
- a delocalisation of economic activity and the associated heterogeneity of economic zones;
- new specialised uses of rural spaces (especially related to tourism), creating new specialised networks of relations in the areas concerned, many of which are no longer localised;
- the fact that people inhabiting rural space increasingly include a diversity of temporary visitors as well as residents;
- the fact that rural spaces now tend to perform functions for non-rural users and can exist independently of the actions of rural people.

This kind of analysis prompted a realisation that there is no longer any such thing as simple rural space, but, rather, a multiplicity of social spaces that overlap the same geographical areas. Another, perhaps less obvious implication is the extent to which conventional rural spaces have been transformed by touristic processes and practices (see, for example, Baerenholdt *et al.* 2004; Crouch 1999). As several authoritative texts have recently confirmed, rural areas have become increasingly significant in the reproduction of tourism over recent years (see, for example, Butler *et al.* 1998; Hall 2005; Roberts 2004; Roberts and Hall 2001). It is equally clear, however, that tourism has become increasingly significant in the reproduction of rurality, particularly in view of the existence of a clearly changing set of relationships between space and society in relation to the countryside.

In this chapter I want to question how perceptions of rural life and landscape are ordered by forms of leisure and tourism through which rural 'attractions' serve up theatres of rurality using spectacle and simulacrum. In particular, I will highlight how tourists are increasingly being presented with opportunities for *creative* performativity in the countryside, and I suggest that perceptions and experiences of rurality are being specifically informed by these creative practices and performances. For some, the switch of emphasis from commodification to performativity might reflect a translation from viewing rurality through a political-economic lens to a more cultural, even non-representational approach. However, one of Richard Munton's many contributions to the development of rural geographies over the years has been to develop a holistic view of the rural, seeking out complementarities of approach and eschewing particular entrenched positionings of theory and concept. In this vein the chapter also explores the possibilities for bringing together notions of commodity and performance in an understanding of the production and consumption of creativity in the contemporary countryside. This approach relies in part on a narrative of the production of rural space that draws on the ideas of Henri Lefebvre. Accordingly, I suggest that particular practices and performances, especially those associated with tourism, are influential in bringing together how rural areas are conceived, lived and perceived. Tourism, then, affords creative performances that characterise not only rural tourism per se but also rurality more generally as a lived space.

The production of rural space

The sociology of knowledge that has underpinned post-Second World War rural geographies can be conveyed in terms as a series of discrete approaches reflecting broader philosophical fashions and trends (Cloke 2005; Marsden 2005). Thus, positivistic and data-driven geographies have been challenged, and some would argue usurped, by the appearance on the scene respectively of political economy, postmodernism and post-structuralism. Although each of these approaches can be portrayed as self-specific, arising from critique of previously held orthodoxies, it might be more fruitful to regard the rural geography 'landscape' as 'palimpsestual', the cartographies of which rely on the different contours of multiple understandings for their mappings of rurality.

A case in point here is the 'move' in rural geography from viewing rural space as a commodity, to understanding rural space as practice. Urry's (2002) argument for the significance of the commodity-form for rural space envisaged rurality as embedded in the meanings conveyed to tourists through the visual components of landscape and nature. Crouch (2005) and others have discussed 'an alternative discourse on rurality' (p. 356) that theorises the dynamic characteristics of the rural through experiences of consuming tourism in the countryside. Thus:

> green politics is considered in terms of its enactment of ideas of the rural; tourist destinations are considered from the perspective of a performance of spaces and cultures not only as ideas or products but as self-actualization, identity work and active reflexive and embodied consumption.
>
> (Crouch 2005: 356)

In this way, the consumer becomes a significant actor in how the rural is constructed and sense is made of it.

Rather than understanding the production of rural space through projection of sign values as somehow distant from the consumption of that space through lived practice, it is possible to chart palimpsestual relations between them. These ideas about how rural space is conceived, lived and practised are illuminated by the writings of Henri Lefebvre (1991) on the production of space. Lefebvre's contribution is to offer a framework through which to render intelligible the qualities of space that are contemporaneously perceptible and imperceptible to the senses (Merrifield 1993, 2000). The resultant 'spatial triad' sketches out three moments that in reality coalesce in fluid and dialectical fashion, but the recognition of which sets the scene for why new forms of practices and performances bring about different relations between tourism and space in rural areas. In short, and leaning heavily on Merrifield's excellent interpretation, Lefebvre's triad is as follows:

- First, *representations of space* indicates that space, as conceptualised and constructed by a range of technocrats and professionals (architects, planners, developers, social scientists, and so on) and their signs, discourses and objectified representations, presents an order that interconnects strongly with prevailing relations of production. Here, then, is space as it is *conceived*, and, as Merrifield (2000: 174) suggests, 'invariably ideology, power and knowledge are embedded in this representation'.
- Second, Lefebvre talks of *representational space*, the space of everyday experience that is shaped by complex symbols and images of the dwellers in and users of that space. Representational space overlays physical space, making symbolic use of its objects, places and landscapes. Here we are dealing with a more elusive experiential realm, space as *lived*, in which

there are continual interventions from conceived space in the form of actors such as planners and tourism developers and managers actively seeking to make sense of how space is lived experientially.

- Third, Lefebvre identifies *spatial practices* as patterns of interaction that glue together society's space by achieving cohesion, continuity and acceptable competences. These are practices through which space is *perceived* and given a performed identity. However, Lefebvre does not suggest any achievement of coherent spatial practices. Rather, such practices represent a fluid and dynamic mediation between conceived space and lived space that at once holds them together, yet keeps them apart.

Lefebvre's 'trialectic' insists that each element of conceived/lived/perceived space is related to and influenced by the others. As such, particular movements of space are unstable and need to be 'embodied with actual flesh and blood culture, with real life relationships and events' (Merrifield 2000: 175). Capitalism will tend to give dominance to the conceived realm, meaning that lived and perceived moments can be of secondary importance to objective abstractions that reduce the significance of both conscious and unconscious lived experience. However, the inevitably incoherent and fractured nature of everyday life and the notion that many performances will be more than representational allow the possibility that new creative practices and performances are capable of affording a release from conceived space as well as a reinforcing of staged meanings that serve to make you know how you feel (Lorimer 2006; see also Dewsbury 2000).

These ideas about the production have been largely eschewed by rural researchers (but see Phillips 2002) until recent work by Halfacree (2005), who connects formal representations of the rural, such as those expressed by capitalist interests, planners, spatial managers and politicians, with how rurality has become commodified in terms of exchange values. Practices of signification and legitimation have been vital to this process, as the vernacular spaces of the rural become variously appropriated symbolically by producers and consumers. This 'social imaginary', however, can itself become subversive as spatial practices and performances are (re)appropriated from the overarching interests of the dominant (Shields 1999). The performance of rural space, then, can be 'staged' by those seeking to commodify the rural experience, but can also subvert any hegemonic notion of what rural practices *should* be like through new ways of making sense of the rural through alternative practices and performances (Edensor 2001). This assemblage of reflexive practice and pre-reflexive performativity points towards new emphases on 'creativity' in how the rural is experienced.

Changing rurality, changing commodity

In my 1993 study I sought to account for the ways in which tourism has been actively involved in the symbolic reconstruction of rural space as well as in its material reconstruction through the production of new sites, facilities and opportunities (Cloke 1993). In recent decades there has been a shift in the nature and pace of commodification in rural Britain as new markets for countryside commodities, and notably those associated with leisure and tourism, have been opened up. The production of rural space has taken new shape through the framing politics of neo-liberalism, and in particular through the specific outcomes of privatisation and deregulation, which have both released rural land for new purposes and created a conception of a multifunctional countryside in which these new rural spaces can take root. Basic ideas such as 'a day out in the country' or 'a rural holiday' have begun to take on new meanings, reflecting opportunities to visit newly commodified 'attractions' in addition to more traditional and less 'pay-as-you-enter' rural pursuits. Consumption of and through these rural attractions has often reflected new forms of old values. Conventional concerns for pastoral idyll, history and heritage, traditions, outdoor pursuits, and the like are still evident. However, such concerns are served up differently through attractions and spectacles that offer new touristic practices to participants while reproducing conventional signs, symbols and displays.

New rural commodity forms, then, suggest changing representations of rural space in that they are shaped and enabled by distinct conceptions of what are *appropriate* new land uses, sites and attractions for the countryside. The needs of changing rural production systems, notably in terms of the necessity for farm diversification, are aligned with new forms of commodity consumption that broadly uphold enduring conceptions of the rural as idyllic, pastoral, close to nature, rich in heritage, safe and problem free, and so on. These representations of rural space are associated with new representational spaces that symbolise and present new images of what it is like to live in that space, telling us how we should feel in and about it. Equally, new spatial practices allow new rural spaces to be perceived and given a performed identity, notably through patterns of interaction involving observing and participation in event-spectacles that somehow perform rurality. Such performances often strive towards some 'authenticity' (however postmodern), but there is also evidence that staged performances of new ruralities can go well beyond the real objects and relations of the sites and buildings concerned. Here, the conspicuous consumption of the symbols on offer indicates that some new rural attractions are emphasising signs that are unrelated to the specific reality of a place, its landscape and its history. In the terms suggested by Best (1989), the society of the spectacle may be nudging up against the society of the simulacrum in which rural commodities are being eclipsed by their sign-values, which can be altogether unrelated to the realities of rural space (Baudrillard 1983).

What this account lacked, however, was a wider appreciation of the growth of so-called cultural consumption (relating, for example, to art, food, fashion, music, and so on) that is fuelling the 'symbolic economy' of particular regions (see Ray 1998; Zukin 1995). In particular, cultural tourism has increasingly sought to engage tourists experientially. As Amin and Thrift (2002) point out, the mere production of goods and services is being overtaken by the production of 'experiences' that draw the consumer into the creative environment of rurality. In particular, there is a significant role for tourists to become 'co-producers' of their own experiences, for example through packages that provide painting, cooking or photography holidays at particular sites (Prentice 2004).

We can therefore begin to appreciate a 'creative turn' in the experiential consumption of rural areas, particularly through tourism (Richards and Wilson 2006). In city contexts the new entertainment economy has been shown to provide new sets of living embodied geographies of creative involvement, representing a 'performative push' to processes of commodification (Amin and Thrift 2002: 125). These creative tourisms provide engaging, interactive experiences that aid personal development and identity formation, and serve as repositories for increased creative capital (Richards 2000; Richards and Raymond 2000). According to Richards and Wilson (2005), creative tourisms are manifest in different forms, of which three stand out:

- *creative spectacles*, such as travelling art exhibitions, that provide creative experiences for passive consumption by tourists;
- *creative enclaves*, such as artists' zones, where populations of culturally creative people attract visitors to visually and emotionally vibrant spaces;
- *creative tourisms*, in which consumers participate in key creative activities and challenges involving active development of particular skills.

While creative spectacles, enclaves and tourisms are by now well documented in urban settings, there has been little indication of their impacts in rural areas (but see Lutyens 2004).

Accepting that many rural areas have redefined themselves as consumption spaces in which the commodification of nature, heritage and tradition has transcended agricultural production as the key signifier of rural space, it is now possible to question what, if any, impacts the 'creative turn' in tourism has exerted on the reproduction of rural space. In terms laid out by Richards and Wilson (2006), the markers of this creative turn reflect a shift from cultural tourism to more skilled forms of consumption, resulting in reformulation of identity and subjectivity and the further acquisition of cultural capital. It is important to trace any such shift towards creativity in other space settings.

Creativity, tourism and the reproduction of rural space

It seems likely that the creative commodification of experiencing rurality will differ considerably from its urban counterpart, reflecting some of the markers laid down by Richards and Wilson (2006) but dressing them in rather different spatial clothing so as to accentuate the distinctiveness of rurality, and its landscapes, natures and heritage. We can speculate, therefore, that creativity in the rural setting will be at least in part prone to the reproduction of idyll-ised constructions of rural space, and the activities that are deemed to be appropriate within that space. After all, rural spectacle will often be based on particular forms of nature–society relations in which people either collude with or pit their wits against the non-human agents of nature (Szerszynski *et al.* 2003). Creative spaces might represent the humble multifunctionality of a village hall rather than the sweeping artistic *quartiers* of metropolitan space. Creative activities might again consist of a humble learning to appreciate local produce rather than more glitzy skills acquisition. Alternatively, creativity in rural tourism may suggest new ways of understanding creativity, for example in terms of hybrid nature–society performances based around eco-experience or adventure. What is clear, however, is that various practices that offer tourists the opportunity to develop their creative potential through forms of active participation in rural contexts are likely in turn to lead to further reproductions of rural space in terms of Lefebvre's triad. In particular, new creative spatial practices will begin to offer potential for gluing space together differently, and providing space with a different performed identity.

In what follows, I draw on research in the United Kingdom (western Scotland and Devon) and to a lesser extent in New Zealand (Kaikoura and Queenstown) to illustrate different strands of how creativity in tourism are associated with the reproduction of rural space. Some examples are much more fully formed than others, and my account is by no means exhaustive. Nevertheless, I suggest four strands of creative performance, recognising that each is neither mutually exclusive nor necessarily mutually compatible in any particular rural place.

Tasting

Here I refer to a range of practices in which tourists (and indeed local people, for I suggest that they too sometimes act as tourists, even when in their home locality) *taste* the creative performance of others, and in so doing develop their own creative potential and expose their identity to change or cultural acquisition. In its simplest form, tasting may reflect attempts to bring cultural events and opportunities into rural space when these previously might only have been accessed in the city. In Devon, for example, the *Nine Days of Art* project is an artist-led initiative that provides a trail around venues throughout rural areas of the county to view jewellery, textiles, ceramics, painting, sculpture, photography and printmaking both inspired by the rural/coastal environments and placed within them, in the creative spaces of homes, studios, galleries, village halls, hotels, country houses and even outdoor locations. The project not only provides residents and tourists with the opportunity to appreciate and purchase rural 'art', but has sparked off several subsidiary packages whereby people can learn to 'do art' for themselves. The *Villages in Action* project, subsidised by local authorities and the Arts Council, brings a programme of performance arts (theatre, comedy, music and dance) to 63 village halls in south Devon. The taste of performance is thereby brought into rural space, where it encourages existing and new performative potentials and credentials.

Tasting, more literally, can involve opportunities to sample and get to know local food and drink, including the production and presentation. Tasting occurs at very different scales; New Zealand's Marlborough region, for example, is becoming sufficiently geared to tourism to suggest a highly organised wine tourism industry. Elsewhere, however, tasting local food and drink forms the basis of more limited spectacle and site-specific attraction. In Scotland, the Mull and Iona Community Trust, for example, organises an annual *Taste of Mull and Iona Food Festival*, bringing together its food and whisky producers, restaurateurs and shopkeepers in a frenzy of events and opportunities for tourists to get into Mull's food culture. The invitation to 'come and share our food with us' is combined with unmistakeable messages about food, place and identity:

> We are more aware than ever before that 'we are what we eat'. We need to know *what* we are eating and *where* it comes from. Food from Mull and Iona is food from farmer to consumer – Real Food – food you can put a face to!

The localised identity of rural food offers tasters the knowledge credentials to be more discerning, sustainable and identity conscious. Other facilities emphasise this point. *Sharpham Vineyard* in Devon invites tourists to 'learn the fine art of making wine and cheese' and, perhaps subconsciously, to learn from their visit about the credentials to be gained from more discerning food and drink choices in the future.

It might be argued that tasting represents a passive form of cultural tourism. Rather, I suggest that it involves practices and performances that develop creative knowledge, intuition, capacity and skill. Interaction with art exposes the rural tourist to signs that both reinforce the rural idyll and contest that idyll with more dystopic interpretations of countryside. Music in the village hall permits the learning and honing of taste, as does food and drink tourism. In so doing, rural areas become replete with creative spaces and practices that are capable of performing the rural differently.

Placing

Another form of creative performance comes with the interaction of rural tourists with imaginative creative performances. For some time now, the representation of rural space has commandeered literary figures and narratives (Heriot Country, Lawrence Country, Robin Hood Country, Lorna Doone's Exmoor, and the like) to convey key significations about particular places, their history and their heritage. By visiting these places, tourists not only may learn more about particular authors and their literary imaginations via interpretive attractions and commodities, but also are enabled to reread the original texts with these 'real' rural settings in mind. Placing imaginative texts is therefore a performative practice that in turn contributes creativity to the place(s) concerned.

The use of rural locations in mass-media programming and filming has added new dimensions to the creative performativity of placing. While this clearly contributes to the conception of some rural spaces as film sets, thereby helping to reconstitute such places as tourist magnets (the village of Goathland in the North York Moors, for example, is the location for the popular television programme *Heartbeat*, and is now the most visited site in the National Park), visitors will inevitably have to engage in imaginative performances to perceive film-space experientially (see Phillips *et al.* 2001). Travellers on Harry Potter's railway in western Scotland have to work hard to connect tourist scene with film scene, as do visitors to New Zealand as *Lord of the Rings* country, especially given the digital enhancement deployed in such movies, which performs radical transformations of rural locations. Such places do, however, offer tourists the opportunity to develop their creative potential through place participation.

A good example of place participation is the recent popularity of Tobermory on Mull in terms of its status as the filming location for the children's television programme *Balamory*. Regular streams of young children (and their parents and grandparents) take the ferry from Oban and the bus to Tobermory so that they can explore the various brightly coloured houses that are used in the programme (Connell 2005). The island's tourism managers have in this case been wary of representing place identity in terms of the imaginative texts associated with Balamory. Although providing visitors with a Balamory map, tourist authorities have avoided elaborate representations of space, for a number of reasons, including the fact that copyright restrictions prevent further commodification of the brand; local people can react negatively to tourist congregations outside private dwellings; and child-oriented day tripping is less valuable economically than tripping by other target tourist groups, possibly conflicting with longer-term target clientele who are attracted by representations of wildness and nature (see below). Nevertheless, the placing of Balamory in Tobermory has created a place spectacle that is performed via a dutiful trudge around a mapped network of televised sites.

While creating – and touristically re-creating – Balamory might be thought of as a relatively undemanding performance of cultural tourism, a series of new projects to place performative arts projects in places of natural beauty are intended to make more serious demands on tourists. Perhaps the best known of these is a Glasgow-based arts organisation (NVA)[1] project combining drama, music, history and night hiking on the Trotternish peninsula of the Isle of Skye. Two hundred participants each night are taken on a torch-lit walk across rugged terrain. Aspects of the landscape are illuminated by lightscapes and soundscapes: 'It ravishes the senses with a fantastic and beautiful night-time journey ... above all, it makes us think about and feel the relationship between humankind and nature' (*Observer*, 10 July 2005: E8). Participating in this 're-placing' of part of Skye offers tourists cultural distinction based on restricted opportunity in an elite landscape. However, this too can be seen as a touristic performance of placing, a creative interaction with imaginative texts in a rural setting.

Performing creatively

The 'creative turn' in tourism brings immediately to mind the upsurge of opportunities for tourists to learn new skills and undertake recognisably creative activities. Rural tourism is now replete with such opportunities. Any self-respecting country house hotel now offers add-on creativities: learn to cook country-style, to fish, to shoot, to ride, to engage in various forms of environmental art, and so on. To some extent these offerings represent the capitalist imperative of conjoining new commodity forms with which to refresh traditional businesses, and while some of these creative activities present obvious connections with traditional conceptions of rural space, other attractions (murder mystery weekends, the opportunity to learn to play tennis, writing or music master classes, etc.) reflect more of a society of the simulacrum than any particular fusion with rural signification.

I want to suggest, however, that offering tourists the opportunity to develop their creative potential will not be restricted to a series of obviously 'country' cultural activities. The creative performance of 'being rural' has moved on in many parts of the developed world. An example of alternative creative performance can be found in the adventure tourism industry of New Zealand (Cloke and Perkins 1998, 2002; Cater and Smith 2002). Here an amalgam of particular circumstances (not least the 'outdoorsiness' of New Zealanders, the technological innovation of jet boats and bungee technology, and the willingness of government to license adventure tourism businesses in elite environmental sites) has resulted first in key brand-leading firms (Shotover Jet and A.J. Hackett), then an explosion of follow-on operations, providing opportunities for adrenaline-fuelled adventurous pursuits in rural areas of New Zealand. Engaging in bungee jumping or white-water rafting may appear simply to be a creative spectacle, with the active involvement of the few providing exciting events for others to watch. Our research suggests, however, that the performance of adventure contributes a number of highly creative aspects to the spatial practices and representations of rural New Zealand, which has become signified and experienced in terms of activities of adventurousness. In particular, participants in adventurous activities report a sense of having to overcome their own fear as well as having to overcome the forces of nature. Not only do they win the cultural credentials of having successfully completed the activity concerned, but in so doing they ask questions of their own identity and often claim to be 'different' after the event. At the same time, the actual performance of, say, bungee jumping defies representation. Post-jump interviews encounter jumpers struggling to put into words what they have just experienced. It is a kind of performativity in which although the actual process is staged, nevertheless the unfolding event is entirely immanent, and resistant to representational signification. In these ways the creative performance of adventure appears to shift the conceived–lived–perceived register of rural space into different directions from those traditionally defined. Although the Queenstown area of New Zealand has fulfilled its self-appointed promise as the 'Adventure Capital of the World', this adventurous motif is being copied into the touristic offerings of many rural areas, including many in the United Kingdom, where practices such as climbing, canyoning and rafting are becoming commonplace.

One further disjuncture from the performance of rural idyllism is also worthy of mention. The post-productivist countryside is beginning to play host to a range of activities and practices that sit somewhat uneasily alongside traditional conceptions of the production of rural space. For example, alongside the farm parks, potteries and farmers' markets of contemporary rurality there are a growing number of *paintball sites* offering opportunities for 'adrenaline-pumping' 'full-on combat'. Although military uses of the countryside are familiar, they are often closed off and always non-participatory. Here, a different form of creativity is emerging where skilled consumption is significant, identity formation is challenged and the cultural capital among particular group cultures is assured, yet in a form that is dystopic to the production of spaces of rural idyll.

Performing interactively

Here, I suggest that the 'creative turn' in rural tourism needs also to encompass how tourists interact with the creativity of nature. Rural studies has in recent years embraced the idea of 'hybrid geographies' of nature–society relations in which relational ensembles of humans, non-humans (whether living or not) discourses, technology, and so on are taken seriously as the formative networks of agency (Whatmore 2002; Cloke and Jones 2001, 2004). This philosophical movement requires nature in its multifarious guises to be recognised as a co-constituent of rural places, and while this is no easy task, given the anthropocentric assumptions of most social science, particular nature–society relations can be used to signify the wider point. Thus, instead of regarding nature as a backcloth to rural tourism, we can begin to ask questions about how nature performs interactively with humanity in tourism, and vice versa.

At the simplest level, creativity can engage with particular facets of nature in the development of new touristic opportunities. An example of this is the *Mull and Iona Tree Festival*, a two-month celebration of trees combining exhibitions, workshops and trails through which visitors can learn of the ancient tree alphabet of the Celts and its associated iconographies. The festival represents an innovative fusion of creative appointments, and fits into the wider calendar of special events to attract tourism to Mull. Another tactic of tourist operators is to upgrade and recommodify classic activities of rural tourism so as to render them more guided, informative and skilled, thus providing them with enhanced credentials. In New Zealand the representation of particularly scenic hikes (or 'tramps') such as the Milford Track and the Routeburn Track as 'Great Walks' has increased the apparent skilled consumption involved, and heightened their popularity. Elsewhere in the country, walks that have always been available have been similarly reproduced. Bell's account of walking up to Fox Glacier near the west coast of New Zealand provides an excellent example:

> The experience, which I'd remembered from previous times as an easy walk, had now been mediated for these tourists by converting the placid (non-paying) traveller to a (paying) alpine adventurer, with all the correct

garments and equipment required for such an adventure. By augmenting (artificially) the scale and danger, in effect they become participants in a recapitulation of nineteenth century experience of the vast and sublime. Literally, the fifteen-minute jog in track stores has been transformed into a half-day guided 'expedition'.

(1996: 42)

These mediative alterations to the tourist experience can be seen simply in terms of recommodification for greater exchange value, but alongside such motivations there is a tacit acceptance that nature itself can perform in articulate ways to challenge, to induce embodied response, to make difficult and so on. In the United Kingdom there has been a similar pattern of branding particular walking activities, for example in terms of the West Highland Way, the Pennine Way and the South-West Coast Path (Wylie 2005). Again, the creative credentials of hiking in the countryside can be enhanced through participation in such branded activities.

In some cases the specific performances of particular creatures are central to the creativity of rural tourism. The burgeoning reputation of Mull as 'the best place in Britain to see wildlife' is underwritten by the iconic performative manoeuvres of eagles soaring, whales surfacing, puffins pausing to stand by their burrows and otters nervously fishing. While tourists can connect with these performances independently, increasing numbers are willing to pay for the services of an experienced guide to lead them on their 'wildlife safari' and to pass on essential skills and knowledges. Perhaps one of the ultimate connections with nature's creative performances can be seen in the astonishing growth of New Zealand's ecotouristic whale-watching and swimming-with-dolphins activities. Research centred on the South Island tourist town of Kaikoura suggests that it is the majestic 'blowing' and 'fluting' of whales and the playful acrobatics of dolphins that provide tourists with magic moments of connection with these cetaceans (Cloke and Perkins 2005). Moreover, such encounters are often presented as moments of unfolding, affording release from the staged and conceived spaces of the tourist industry. In a very real sense, then, the embodied performances of cetaceans lie at the heart of Kaikoura's growth as tourist town. Some tourists are fortunate enough to achieve creative interactions with cetaceans in the wild. For others, just visiting the town and soaking up the representational space signifying the performance of these creatures is enough. In assessing the creativity of cultural tourism in rural areas we do well to recognise the co-constitutive hybridities of human and animal performances. Such contact with the creative performances of nature is becoming increasingly available in the rural areas of the United Kingdom.

Conclusion

In one sense the significance of creativity in rural tourism can be represented in terms of heralding a new turn in the touristic commodification of rural areas. However, when set against Lefebvre's ideas about the production of space, the breakout of creativity needs to be assessed in terms of any resultant practices that are capable of changing the ways in which rural space is brought into being performatively. The research reported here suggests that care needs to be exercised over the potential conflation of rurality and creativity; creativity is multi-faceted and reflects far more about rural space than just the 'obvious' connections with 'rural' cultural activities and skills.

In general, rural space is increasingly *conceived* as a commodity forum, in which rurality is reproduced both as an object of desire and as a stage on which to perform. Much of the apparent creativity in rural tourism deepens the relationship with rurality, and therefore deepens its desire as a place of performance. Some performances enhance the traditional appeal of the rural; others appear to reconceive rurality as a new space for adventure, entertainment, spectacle and the like. Equally, rural space is increasingly *lived* in terms of new portfolios of symbols and images by which dwellers and users experience and make sense of the rural. Many of these messages signify the apparent need for a greater cultural engagement with rurality, involving new skills of observation, understanding and embodied participation. Thus, rurality is becoming signed as a place in which to perform (although performances range across different degrees of passivity and activity), and the creativity *of* the rural emerges through creativity *in* the rural. The *perceived* space of the rural thereby begins to demonstrate new cohesions played out in terms of practices that provide opportunities for visitors to dwell, to reside (albeit temporarily) and to be involved. Involvement ranges from a post-tourist sense of dealing uneasily with the inauthentic, to a seeking out of new elements of the 'authentic', but throughout this range, new performed identities are exhibited, some of which are more than representational in their affordance of release from orthodox conceived space.

Thus, it is clear that creativity in rural tourism is resulting in a range of different practices and performative spaces in which the identity and subjectivity of the tourist can be reformed and enhanced. In some cases, for example in interactions with particular aspects of nature, these practices have the capacity to reinforce perceptions of how traditionally conceived and lived space is played out in rural settings. Alternatively, new creative practices may be regarded as dystopic, and can certainly present conflicting social demands on a particular geographical space, as the example of Mull illustrates. In these latter cases, creative touristic performance does pose serious questions about the production of rural space. New ways of bringing together how rurality is conceived and how it is lived out can involve a release of new kinds of perceptions, as well as a retrenchment of existing ways of being made how to feel. It is in the thrilling and exacting credentials of adventurous and environmentally interactive 'rurals', as well as in the skilled consumption of tradition and heritage, that the creativity of rural space is being discovered in contemporary tourism.

Note

1 NVA is *nacionale vitae activa*, denoting the right to influence public affairs.

References

Amin, A. and Thrift, N. (2002) *Cities: Reimagining the Urban*, Cambridge: Polity Press.

Baerenholdt, J., Haldrup, M., Larsen, J. and Urry, J. (2004) *Performing Tourist Places*, Aldershot: Ashgate.

Baudrillard, J. (1983) *Simulations*, London: Routledge.

Bell, C. (1996) *Inventing New Zealand*, Auckland: Penguin.

Bell, M. (2005) 'Variations on the rural idyll', in P. Cloke, T. Marsden and P. Mooney (eds) *Handbook of Rural Studies*, London: Sage, pp. 149–60.

Best, S. (1989) 'The commodification of reality and the reality of commodification: Jean Baudrillard and post-modernism', *Current Perspectives in Social Theory*, vol. 19, 23–51.

Butler, R., Hall, M. and Jenkins, J. (eds) (1998) *Tourism and Recreation in Rural Areas*, Chichester: Wiley.

Cater, C. and Smith, L. (2003) 'New country visions: adventurous bodies in rural tourism' in P. Cloke (ed.) *Country Visions*, Harlow: Pearson, pp. 195–217.

Cloke, P. (1993) 'The countryside as commodity: new rural spaces for leisure', in S. Glyptis (ed.) *Leisure and the Environment*, London: Belhaven, pp. 53–67.

Cloke, P. (2005) 'Conceptualising rurality', in P. Cloke, T. Marsden and P. Mooney (eds) *Handbook of Rural Studies*, London: Sage, pp. 18–28.

Cloke, P. and Jones, O. (2001) 'Dwelling, place and landscape: an orchard in Somerset', *Environment and Planning A*, vol. 33, 649–66.

Cloke, P. and Jones, O. (2004) 'Turning in the graveyard: trees and the hybrid geographies of dwelling, monitoring and resistance in a Bristol cemetery', *Cultural Geographies*, vol. 11, 180–208.

Cloke, P. and Perkins, H. (1998) ' "Cracking the canyon with the Awesome Foursome": representations of adventure tourism in New Zealand', *Environment and Planning D: Society and Space*, vol. 16, 185–218.

Cloke, P. and Perkins, H. (2002) 'Commodification and adventure in New Zealand tourism', *Current Issues in Tourism*, vol. 5, 521–49.

Cloke, P. and Perkins, H. (2005) 'Cetacean performance and tourism in Kaikoura, New Zealand', *Environment and Planning D: Society and Space*, vol. 23, 903–24.

Connell, J. (2005) 'Toddlers, tourism and Tobermory: destination marketing issues and television-induced tourism', *Tourism Management*, vol. 26, 763–76.

Crouch, D. (ed.) (1999) *Leisure/Tourism Geographies: Practices and Geographical Knowledge*, London: Routledge.

Crouch, D. (2005) 'Tourism, consumption and rurality', in P. Cloke, T. Marsden and P. Mooney (eds) *Handbook of Rural Studies*, London: Sage, pp. 355–64.

Dewsbury, J.-D. (2000) 'Performativity and the event: enacting a philosophy of difference', *Environment and Planning D: Society and Space*, vol. 18, 473–97.

Edensor, T. (2001) 'Performing tourism, staging tourism', *Tourist Studies*, vol. 1, 59–82.

Glyptis, S. (ed.) (1993) *Leisure and the Environment*, London: Belhaven.

Halfacree, K. (2005) 'Rural space: constructing a three-fold architecture', in P. Cloke, T.K. Marsden and P. Mooney (eds) *Handbook of Rural Studies*, London: Sage, pp. 44–62.

Hall, D. (ed.) (2005) *Rural Tourism and Sustainable Business*, Clevedon, Somerset: Channel View Books.

Lefebvre, H. (1991) *The Production of Space*, Oxford: Blackwell.

Lorimer, H. (2006) 'Herding memories of humans and animals', *Environment and Planning D: Society and Space*, vol. 24, 497–518.

Lutyens, D. (2004) 'Bright young Fins', *Observer Magazine*, 7 March, 46–9.

Marsden, T.K. (2005) 'Pathways to a sociology of rural knowledge', in P. Cloke, T.K. Marsden and P. Mooney (eds) *Handbook of Rural Studies*, London: Sage, pp. 3–17.

Merrifield, A. (1993) 'Place and space: a Lefebvrian reconciliation', *Transactions of the Institute of British Geographers*, vol. 18, 516–31.

Merrifield, A. (2000) 'Henri Lefebvre: a socialist in space', in M. Craig and N. Thrift (eds) *Thinking Space*, London: Routledge, pp. 167–82.

Mormont, M. (1990) 'Who is rural? Or, how to be rural: a sociology of the rural', in T.K. Marsden, P. Lowe and S. Whatmore (eds) *Rural Restructuring*, London: David Fulton, pp. 21–44.

Phillips, M. (2002) 'The production, symbolization and socialization of gentrification: impressions from two Berkshire villages', *Transactions of the Institute of British Geographers*, vol. 27, 282–308.

Phillips, M., Fish, R. and Agg, J. (2001) 'Putting together ruralities: towards a symbolic analysis of rurality in the Bristol mass media', *Journal of Rural Studies*, vol. 17, 1–27.

Prentice, R. (2004) 'Tourist motivation and typologies', in A. Lew, M. Hall and A. Williams (eds) *A Companion to Tourism*, Oxford: Blackwell, pp. 261–79.

Ray, C. (1998) 'Culture, intellectual property and territorial rural development', *Sociologia Ruralis*, vol. 38, 3–20.

Richards, G. (2000) 'World culture and heritage and tourism', *Tourism Recreation Research*, vol. 25, 9–18.

Richards, G. and Raymond, C. (2000) 'Creative tourism', *ATLAS News*, vol. 23, 16–20.

Richards, G. and Wilson, J. (2006) 'Developing creativity in tourist experiences: a solution to the social reproduction of culture', *Tourism Management*, vol. 27, 1209–23.

Roberts, L. (2004) *New Directions in Rural Tourism*, Aldershot: Ashgate.

Roberts, L. and Hall, C. (2001) *Rural Tourism and Recreation*, Wallingford: CABI.

Shields, R. (1999) *Lefebvre, Love and Struggle: Spatial Dialectics*, London: Routledge.

Szerszynski, B., Heim, W. and Waterton, C. (eds) (2003) *Nature Performed: Environment, Culture and Performance*, London: Sage.

Urry, J. (2002) *The Tourist Gaze*, 2nd edn, London: Sage.

Whatmore, S. (2002) *Hybrid Geographies: Natures Cultures Spaces*, London: Sage.

Wylie, J. (2005) 'A single day's walking: narrating self and landscape on the South West coast path', *Transactions of the Institute of British Geographers*, vol. 30, 234–47.

Zukin, S. (1995) *The Culture of Cities*, Oxford: Blackwell.

7   Nature conservation in towns and cities

David Goode

Introduction

During the past 20 years the conservation of urban nature has taken its place as an integral part of nature conservation in the United Kingdom. A relative late-comer on the scene, it has provided a new set of objectives that are very different from the more traditional approach to conservation, with its emphasis on the protection of key sites and rare species.

Urban nature conservation differs in a number of ways. It is particularly characterized by an emphasis on the availability of nature to townspeople in the places where they live or work. Access to nature within the urban environment is thus one of its key objectives, and it matters little whether that nature is special or ordinary, so long as it provides an experience of nature close to home. The mix of habitats occurring in British towns and cities includes some which are vestiges of semi-natural habitats that have become encapsulated within the urban sprawl, but which have their origins in the former countryside. All too often such habitats have suffered from the effects of urbanization, losing many of the species that would characterize them in their more natural surroundings. Yet these habitats have considerable value in the urban context as they are frequently the only ones available for people to enjoy.

A second value stems from the special characteristics of urban nature, resulting from colonization of post-industrial landscapes. Many urban habitats result from nature taking over from humankind's previous activities, colonizing disused and derelict land, converting 'brownfield' land into new green oases. Because of this, many such areas have a special quality, an indefinable urban-ness that sets them apart from their country cousins. Orchids growing on pulverized fuel ash, the legacy of former industrial processes, now provide a valuable example of nature in the midst of urban Lancashire, and in so doing provide a link with the area's recent past, the ecological equivalent of industrial archaeology. Such examples are common-place in British towns and cities that owe their existence to the industrial revolution.

Third is the functional importance of natural areas within the urban environment. The value of green space for climatic amelioration and for reducing the risk of flooding has become increasingly appreciated, particularly as a result of recent analyses of the potential impacts of climate change.

A fourth feature of urban nature conservation that sets it apart from the more traditional approach to nature conservation is the strong emphasis on ecological restoration and habitat creation, rather than simply protecting key examples of habitats and species. This is reflected in the creation of new habitats ranging from tiny patches on rooftops and in school grounds, through numerous examples of ecological parks, to the extensive London Wetland Centre covering the former Barn Elms reservoirs. They are all part of a process that aims to encourage wildlife and nature within the urban scene through ecologically sensitive design, often as a part of urban development.

Finally, urban nature conservation differs in the processes it employs. It is in many ways a grass-roots movement, involving and depending on local communities. Urban wildlife groups tend to involve a broader constituency of local people than the more traditional county wildlife trusts, many of which still retain a strong emphasis on special sites and rare species. Urban groups tend to focus on links with people; whether for recording stag beetles, protecting an area under threat, or for popularizing the interest of local sites, it is the people that count. It is these links with the local community that give it its strength.

In almost all its features, urban nature conservation has been at odds with national policies for biodiversity conservation. The national hierarchy of sites for protection places greatest emphasis on those of international and national importance, followed by those of regional or local value. Hardly any of the many important biodiversity sites in towns and cities qualify as Sites of Special Scientific Interest (SSSI) in the national context. Sites of value in the urban environment barely register on the national scale of values, and this is reflected by the fact that only 2 per cent of SSSIs in England are in urban areas.

There are a number of reasons why this is so. In the first case, urban habitats tend to be undervalued, as they commonly fall outside the categories of habitat recognized nationally as being of nature conservation importance. The physical characteristics and processes of urban areas frequently result in a range of habitats and species of particular urban character that are not generally found in the countryside. Some are virtually restricted to specialized conditions resulting from post-industrial landscapes that are poorly described in UK habitat and vegetation classifications. As a result, national policies for biodiversity conservation do not at present give adequate recognition to these specialist urban habitats. A few outstanding examples have achieved SSSI designation, as in the case of invertebrate populations at Canvey Wick in Essex, but these are exceptional cases and there are no clear guidelines for evaluating such habitats.

Although the value of biodiversity for people living in urban areas has become widely appreciated and is now an accepted element of nature conservation policy at national level, this does not sit easily with a national biodiversity strategy that gives priority for protection to habitats and species of international and national importance. Most semi-natural habitats that have survived in urban areas are so heavily modified that they are poor examples when viewed strictly in terms of their intrinsic scientific interest. Yet there are other criteria, reflecting their social and cultural values, that need to be taken into account in the urban context. Protection of important urban habitats requires these other criteria to be recognized.

The same argument applies in the case of ecosystem functions and services. Habitats within the urban environment may have significant value in the range of ecological services that they provide, especially in terms of flood alleviation and local climatic amelioration. While the importance of these functions is recognized in government policies for sustainable development, the hierarchy for protection of biodiversity in planning (based on international and national importance) takes precedence, and as a result the planning process does not give sufficient weight to the functional value of urban habitats. So, in most towns and cities, features such as streams and river valleys, which have a vital role in maintaining sustainable urban drainage, may have no designation for protection, or are at best regarded as having only local value in terms of their biodiversity. The fact is that urban nature conservation has developed a separate rationale that needs to be treated differently from its rural counterparts. This chapter explores some of the issues involved and offers some possible ways forward.

Conserving urban nature through planning

Programmes for the conservation of wildlife sites in urban areas only really started with the formation of the metropolitan counties in the 1970s, when it was first recognized that special provision needed to be made for the protection of such areas. This new development followed publication of a description of the ecology of the West Midlands by the Nature Conservancy Council (Teagle 1978), which drew attention to the vast array of high-quality wildlife sites surviving amid the industrial dereliction of the Birmingham conurbation. A series of strategies was developed in the early 1980s by the metropolitan counties, including London, Manchester, Merseyside, Tyne and Wear, and the West Midlands (see Goode 1989). Other cities quickly followed, with local authorities and urban wildlife groups taking the lead to develop major programmes in Bristol, Edinburgh, Leicester and Sheffield. Most towns and cities now have policies for the protection of urban wildlife sites, which is achieved largely through the designation of Sites of Importance for Nature Conservation (SINCs) by local planning authorities. Some of these programmes have been remarkably successful in ensuring the protection of sites through the planning process despite the absence of a statutory basis for such sites.

The strategy for London, implemented by London boroughs through the London Ecology Unit from 1986 to 2000, was particularly effective (Goode 2005) and now forms the basis of the Biodiversity Strategy for London (Mayor of London 2002), which is an integral part of the statutory London Plan. Other cities adopted similar strategies, supported by specialist ecological teams, as in Merseyside and Manchester, while others have developed arrangements with urban wildlife groups, as in Bristol and Sheffield.

The most successful schemes are those that have been integrated firmly into local government arrangements, through voluntary joint committees representing district planning authorities, as in the case of London and Manchester. These have developed clear strategies for protection of important sites through the planning process. To be effective and successful, such schemes require the following:

- comprehensive survey of habitats throughout the urban area concerned;
- evaluation of sites to determine their nature-conservation importance based on criteria relevant to the urban context, including social and educational criteria;
- implementation of the scheme through strategic planning and development control, with strong policies for site protection in local development plans;
- expert ecological advice available to planners on the implications of individual development proposals and to support the local authority in the event of planning appeals.

Such an approach for implementation and protection of Sites of Importance for Nature Conservation (SINCs) has been more robust and effective than is generally the case where planning authorities depend on local wildlife trusts to provide the necessary ecological input. Although many towns and cities have developed urban nature conservation strategies, there has been a varied degree of success. Some of the possible reasons for failure are as follows:

- the absence of a legislative requirement for development of such strategies by local authorities;
- failure to undertake comprehensive ecological surveys and to update them regularly;
- inadequate long-term funding to maintain local biological records centres, particularly site-based information systems for use by local planning authorities;
- lack of an agreed approach regarding appropriate criteria for conservation evaluation in urban areas, especially the criteria necessary to assess the social or educational value of sites;
- absence of in-house ecological expertise to meet the needs of planning authorities, especially at district level.

However, the situation has changed radically with the publication by the Department for Environment, Food and Rural Affairs (DEFRA 2006) of its guidance on the identification, selection and management of Local Sites (DEFRA 2006) which should assist in resolving some of these issues. This states:

Local Authorities should provide leadership in establishing and maintaining partnerships and systems to identify and manage Local Sites. It may be more effective for the County or Unitary authority to take the overall strategic lead as other relevant frameworks such as Local Record Centres and

Biodiversity Action Plans are usually structured at this level. In many areas it may be more effective for Local Authorities to collaborate in partnerships to run a single unified Local Sites system over their combined administrative areas, particularly in tiered county/districts.

This guidance builds on existing good practice and provides the basis for selection and protection of local sites, recognizing their importance as a component of the national series of biodiversity sites. It gives recognition to the value of such sites in urban areas, stating: 'In populous areas that are poorer in high quality natural environment, sites of lesser intrinsic ecological interest may still be of substantive nature conservation value for the opportunities they provide for the appreciation of nature'. It also specifically recognizes the value of such sites for environmental education.

The DEFRA guidance goes a long way to formalizing what has already become accepted practice in many towns and cities where local authorities have taken the lead in developing programmes for the protection of such sites. The guidance states:

> All Local Sites systems should have a set of clear and locally defined site selection criteria with measurable thresholds developed with reference to the standard set of criteria listed below. Some or all of these can inform the development of individual measurable thresholds for Local Site selection criteria, providing a structured and systematic approach to the description and assessment of sites.

The criteria are as follows:

- size or extent;
- diversity;
- naturalness;
- rare or exceptional feature;
- fragility;
- typicalness;
- recorded history and cultural associations;
- connectivity within the landscape;
- value for appreciating nature;
- value for learning.

The last four of these criteria are particularly important in the context of urban areas. *Recorded history and cultural associations* allows for recognition of local historical features, which may include industrial archaeology or other aspects of local history that have a bearing on the present-day ecology. *Connectivity* is important in providing linkages through the urban environment. This is particularly important for protection of the green infrastructure, especially in terms of urban streams and rivers, but also for provision of recreational cycle and walking routes linking local sites. *Value for appreciating nature* and *value for learning* are of paramount importance in the urban context, allowing for protection of a range of sites of truly local importance which together provide a national resource of immense value. The inclusion of these four criteria in national policies will assist considerably in the development and implementation of programmes for urban nature conservation.

Guidance from DEFRA also advocates the need for Local Sites Partnerships, drawing on the many organizations that have a role to play in this field. The role of the Local Sites Partnership should be to do the following:

- agree the basis for site selection;
- coordinate site selection procedures including survey and identification of candidate sites;
- actively promote site management;
- coordinate funding provision;
- promote educational use where appropriate;
- establish a process for monitoring the condition of the selected sites;
- review the operation of the local sites system at suitable intervals;
- promote the role and importance of local sites at a strategic level;
- promote the enhancement of sites through buffering and increased connectivity.

The DEFRA guidance also suggests that community strategies can provide a useful framework for adopting local sites, and that such sites must be submitted to the local authority for inclusion within its Local Development Framework at the earliest opportunity. It goes on to say that while local sites will not be individually identified within the Regional Spatial Strategy, partnerships might make representations to their regional planning bodies to ensure that the role of local sites is recognized as part of broader regional policy addressing biodiversity.

This guidance goes a long way towards bridging the gap between national biodiversity policies and the needs of urban nature conservation. The government's Planning Policy Statement 9, which sets out national planning policy for biodiversity, recognizes that local sites have a fundamental role to play in meeting national biodiversity targets, contributing to the life and well-being of the community and in supporting research and education. Planning Policy Statement 12 on Local Development Frameworks states that these should identify all local nature conservation areas on proposals maps and that criteria-based policies should be established in development plan documents against which proposals for any development on or affecting such sites will be judged. The recent guidance (DEFRA 2006) now provides the basis for such schemes to be implemented.

The conservation of biodiversity as part of a broader sustainability agenda is addressed by the Town and Country Planning Association in *Biodiversity by Design: a Guide for Sustainable Communities* (2004). This recognizes the function of natural areas and other green space in towns and cities in terms of climatic regulation, flood alleviation and health. It provides numerous case studies demonstrating the value of the ecological services provided by such areas, and demonstrating how biodiversity can be built into design at every level from master planning down to provision of 'green roofs'. The value of the green infrastructure deserves far greater recognition in government guidance on urban development. There is need for urgent action to provide guidance on provision of green space and biodiversity along the lines advocated by the Town and Country Planning Association (2004).

The national biodiversity action plan

The national action plan for biodiversity (Department of the Environment 1995), implemented in response to proposals made at the Rio Earth Summit of 1992, has also had a considerable impact on the development of urban nature conservation. One of the key features of the UK Action Plan is the production of Local Biodiversity Action Plans (LBAPs). These were aimed not only at ensuring delivery of the overall plan at the local level, but also at providing a mechanism for engaging a wider constituency in biodiversity conservation, and catering for local rather than national needs (see Local Issues Advisory Group 1997). This has resulted in the production of such plans for most urban areas, with detailed proposals for the conservation of typically urban species such as bats, swift, house martin, black redstart, song thrush and house sparrow. These local action plans have also defined priority habitats for conservation in each local area, with habitat action plans included for parks, cemeteries, private gardens, wastelands and even docklands. Most towns and cities have an ongoing LBAP reflecting local priorities.

The primary functions of LBAPs, set out in guidance issued by the UK Local Issues Advisory Group of the UK BAP in 1997, are as follows:

- to ensure that national targets for species and habitats, as specified in the UK Action Plan, are translated into effective local action;
- to identify targets for species and habitats appropriate to the local area, and reflecting the values of people locally;
- to develop effective local partnerships to ensure that programmes for biodiversity conservation are maintained in the long term;
- to raise awareness of the need for biodiversity conservation in the local context;
- to ensure that opportunities for conservation and enhancement of the whole biodiversity resource are fully considered;
- to provide a basis for monitoring progress in biodiversity conservation at both local and national level.

There is no doubt that these LBAP's have been extremely effective in bringing biodiversity conservation firmly into the work of many urban local authorities.

The government's announcement on the rationalization of local authority plans identifies the LBAP as one of the plans to be subsumed into Community Strategies (Office of the Deputy Prime Minister 2002). Local authorities are expected to demonstrate that local biodiversity planning has been considered within their community strategy and that such strategies as a whole are informed by the purposes of biodiversity planning.

One of the key features of LBAPs is the development of action plan partnerships, which include a wide spectrum of players all of whom are involved to a greater or lesser extent in delivering the plan. For some of these bodies, like the urban wildlife trusts, conservation is their mainstream activity. They are directly involved in the protection and management of nature reserves, and in developing programmes involving local people. For others, like the British Trust for Community Volunteers (BTCV) and local Groundwork Trusts, it is a large part of their core activities, but there are other bodies involved that deal with biodiversity only as a part of their overall operations. These include water companies, Network Rail, port authorities, British Waterways and regional health authorities. In most cases, these local partnerships are led by the local authority. A key to the success of local action plans has been to engage with all these players, and in some cases to ensure that they sign up to a biodiversity charter. Many of these existing partnerships will readily provide the basis for local sites partnerships, as advocated by DEFRA (2006). LBAPs are also well integrated with issues at the local level in towns and cities, and are particularly compatible with the whole philosophy of urban conservation.

In one respect the UK Biodiversity Action Plan (BAP) has failed to address the issue of urban habitats effectively. Although it aims to encompass all aspects of biodiversity in the United Kingdom, including towns and cities, through the setting of national targets for priority habitats and species, it has failed to do this for urban habitats. A number of broad habitat types (such as woodland) are recognized, and within each type priority habitats are identified. Urban habitats have, however, posed something of a problem. Although a broad urban category was recognized from the outset, because of the lack of a recognized classification of urban habitats, priority habitats have yet to be defined. However, a review of the habitat and species coverage of the UK BAP is currently under way and has provided an opportunity to assess the coverage of habitats and associated species which can be regarded as characteristically 'urban'. A recent study, commissioned by English Nature (Tucker *et al.* 2005), concluded, *inter alia*:

- There is a strong case for the separate recognition of a group of habitats under a combined category of 'Post-industrial sites of high ecological quality' as a Priority Habitat. These habitats are both ecologically valuable and threatened.
- The majority of urban habitats are of moderate or high overall conservation importance, often in terms of both biodiversity and social value.
- Conservation efforts should not solely focus on priority habitats and priority species, but should also aim to maintain all remaining semi-natural habitats

in urban landscapes and strive to enhance their ecological quality and connectivity.

- The overall number of priority UKBAP species associated with urban habitats has previously been underestimated.
- The report also recognized the need for further research regarding urban habitats and species and recommended specific habitat types which need to be addressed.

The UK Inter-Agency Urban Habitat Working Group broadly endorsed these recommendations, and the Urban Group of the England Biodiversity Strategy has recommended that a new Priority Habitat of 'Post-industrial land of high ecological value' should now be included.

Access to nature

In 1992, English Nature promoted a study to provide targets and guidelines for providing adequate accessible natural green space in urban areas. The result suggested two minimum targets (Box and Harrison 1993):

- an urban resident should be able to enter a natural green space of at least two hectares within 0.5 kilometre of their home;
- provision should be made for local nature reserves in every urban area at the minimum level of one hectare per thousand population.

Following this, a further study recognized that accessible natural green spaces do not have to be as large as two hectares, but to be accessible they do have to be in the right place, within five minutes' walking distance of home, and they have to be places where people feel safe (Harrison *et al.* 1995). On the basis of this work, English Nature promoted the use of this model for provision of accessible natural green space, recognizing that such areas contribute much to people's health and well-being, and thereby contribute to sustainable communities.

A further report reviewed progress and made recommendations for implementation by local authorities (Handley *et al.* 2003). The project found that awareness of the model was very low among the local authorities surveyed, and recommended that for progress to be made, local authorities would need to incorporate it as an element in development of their green-space strategies. The Greater London Authority is one of the few that has actually implemented an approach of this kind, but in this case the distance criterion has been set at one kilometre. Details of the approach adopted in London are provided in the next section.

The London biodiversity strategy: a working model of urban conservation

Over the past 20 years an innovative programme for nature conservation has been developed and implemented in London. Its purpose is to protect and enhance London's natural areas and their associated species, and to make it possible for Londoners to have greater contact with nature in their everyday lives. The programme has involved many different players, including official agencies, especially the London boroughs, and voluntary bodies such as the London Wildlife Trust, often strongly supported by public opinion, particularly at the local level. New approaches with a strong social dimension that were at first seen as a radical departure from traditional nature conservation have since been adopted as an integral part of city management. The overall programme has now been adopted by the Mayor of London as the basis for his Biodiversity Strategy, which forms part of the statutory London Plan (see Goode 2005).

Identifying important sites

When the Greater London Council took the first steps in developing a nature conservation strategy in 1982, knowledge of London's ecology was patchy and incomplete. The first priority, therefore, was to undertake a comprehensive survey of wildlife habitats. The strategy required detailed ecological information for all places of potential significance, including information on the kinds of habitats and an assessment of their importance. Priority was given to areas of open land of potential significance for nature conservation. Formal parks and cemeteries, private gardens, playing fields and open areas with little wildlife interest, such as arable land, were all excluded from the survey. An initial desk study using air photography resulted in over 1,800 'sites' being selected for survey, totalling about 20 per cent of the land area of Greater London. For each site, information was collected on the types of habitat and dominant species, overall richness of plant species, presence of rare or unusual species, current land use and accessibility.

This survey provided the starting point for selection of Sites of Importance for Nature Conservation in London. A standardized set of criteria were used for comparing and evaluating sites. Although many of these criteria are similar to those developed by UK government agencies for selecting sites of national importance (such as species richness, size and presence of rare species), there are some essential differences. Public access and value for environmental education are examples (Goode 1999).

The data have since been periodically updated through more detailed surveys of each individual London borough. Since 2000 the Greater London Authority (GLA) has implemented a rolling programme of resurvey covering about three boroughs each year, with the intention that all 33 will be covered within ten years. Over the past 20 years the database has provided a vital tool in strategic planning and for advising on the ecological implications of proposed new developments. Probably the most detailed ecological database of any part of the United Kingdom, it now provides essential information for implementation of the Mayor's biodiversity strategy for the capital.

The procedure for selecting sites for protection was first described in *Nature Conservation Guidelines for London* (Greater London Council 1985). This contained a series of ecological policies for use in strategic planning in London and set out the rationale to be used for deciding which areas are important for nature conservation. It provided the basis for the system used by the London Ecology Unit from 1986 to 2000. Although some changes have occurred in the detailed approach, the rationale remains much the same as that developed in 1985, and it has been widely accepted as the basis for nature conservation planning in London. Although at that time non-statutory, it was endorsed by the London Planning Advisory Committee for use in unitary development plans for all London boroughs in 1995. The same policy, criteria and procedures for identifying nature conservation sites were adopted by the Mayor of London in 2000, and are set out in full in his Biodiversity Strategy (Mayor of London 2002).

Categories of protected site

The strategy is based on a hierarchy of sites at three levels: London-wide, Borough and Local. Those of London-wide strategic significance are called Sites of Metropolitan Importance for Nature Conservation. They include nationally protected sites, such as National Nature Reserves and Sites of Special Scientific Interest, together with many other sites that together represent the full range of habitats in London. The second category comprises sites of significance to individual London boroughs, and a third category, that of local sites, involves those important at neighbourhood level.

The use of these three different levels of importance is an attempt not only to protect the best sites in London but also to provide each area of London with accessible wildlife sites so that people are able to have access to nature within their local neighbourhood. This hierarchy means that sites of London-wide importance are chosen in the context of the geographical area of Greater London. Sites of borough importance are selected from the range of sites in each individual borough. Sites of local importance are those that are valued by local residents, schools or community groups at the neighbourhood level.

At the London-wide level about 140 Sites of Metropolitan Importance are identified. They are distributed throughout London and vary in size from only a few hectares to over 1,000 hectares. Most (90 sites) are less than 100 hectares, of which 55 are less than 50 hectares apiece. A few Sites of Metropolitan Importance have been lost to development since the London Ecology Committee first endorsed the list in 1988. Most of these were wasteland sites that were already scheduled for development. Additional sites have been added to the list over the years as individual boroughs have been surveyed in greater detail. The Mayor endorsed the list of Metropolitan Sites in 2002, and these sites are given statutory protection by policies in his statutory London Plan, which now provides the strategic planning framework for London (Mayor of London, 2004).

As a result of the detailed surveys for individual boroughs, the overall strategy for London identifies over 1,500 sites that are designated as being of importance for biodiversity conservation in borough plans. This total includes all three categories of protection, namely metropolitan, borough and local. A significant number of sites designated through this process are also protected as statutory Local Nature Reserves (LNRs). This is a designation made by the boroughs to give a greater degree of protection in the long term and to promote greater public access. A total of 105 areas are now designated by London boroughs as LNRs, compared with only two in 1980.

Areas of deficiency

Those parts of London that do not have good access to high-quality wildlife sites are identified as areas of deficiency in access to nature (Figure 7.1). These are defined as built-up areas more than one kilometre from an accessible metropolitan or borough site. The Greater London Council first made proposals for such areas as a critical element of the newly developing nature conservation strategy (Greater London Council 1985). Detailed maps defining such areas of deficiency have now been produced for all London boroughs. These maps assist boroughs in identifying priority areas for provision of new habitats and aid the choice of Sites of Local Importance. Local sites are chosen as the best available to alleviate this deficiency. Such sites need not lie in the area of deficiency, but should be as near to it as possible. Where no such sites are available, opportunities should be taken to provide them by habitat enhancement or creation, by acquisition of land capable of fulfilling this function, or by negotiating access and management agreements and improving access routes in the surrounding urban area.

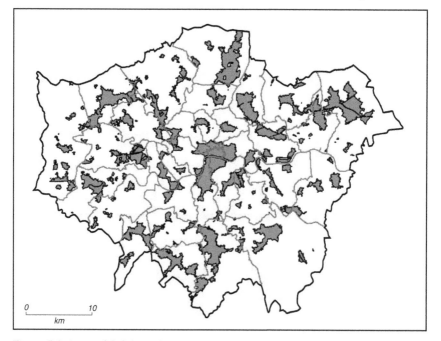

Figure 7.1 Areas of deficiency in access to nature, Greater London Authority, September 2006.

Recognition of such areas of deficiency goes some way to meeting the proposals made by Box and Harrison (1993) for provision of accessible natural green space. The approach adopted in London remains one of the few examples of such a scheme actually implemented as part of regional planning.

Green corridors

Green corridors are also recognized in the planning process. These are relatively continuous areas of open space leading through the built environment, which may link sites to each other and to the green belt and metropolitan open land. They may consist of railway embankments and cuttings, roadside verges, canals, parks, playing fields, cemeteries and river valleys. Although some may not be accessible, many such corridors, especially alongside waterways, do have public access and provide important greenway links for recreation. These green corridors also contribute significantly to the functional ecology of the urban environment.

Implementing the strategy through strategic planning

The Mayor's Biodiversity Strategy, published in 2002, includes specific policies and proposals to protect and enhance biodiversity through strategic planning. These policies are also contained in the statutory London Plan (Mayor of London 2004). Protection of Sites of Importance for Nature Conservation is covered by two major proposals, as follows:

* *Proposal 1.* The Mayor will identify Sites of Metropolitan Importance for Nature Conservation. Boroughs should give strong protection to these sites in their unitary development plans. The metropolitan sites include all sites of national or international importance for biodiversity.
* *Proposal 2.* Boroughs should use the procedures adopted by the Mayor to identify and protect Sites of Borough and Local Importance for Nature Conservation. The Mayor will assist and advise them in this.

The effect of these proposals is that the hierarchy of designations in London is now subject to statutory planning procedures. But it goes further than that, since the Biodiversity Strategy states that the Mayor will measure the success of his strategy primarily against two targets: first, to ensure that there is no net loss of Sites of Importance for Nature Conservation; and second, that the areas of deficiency in accessible wildlife sites are reduced. Monitoring of these targets is addressed in *The Mayor's State of the Environment Report* (Mayor of London 2003).

Implementation of the Mayor's policies to reduce such areas of deficiency is achieved through individual London boroughs as planning authorities. Areas of deficiency are identified in the five sub-regional frameworks that form part of the overall London Plan. Individual boroughs are asked to identify areas of deficiency in access to nature and to address these in their local development frameworks through three processes:

- the natural value of an accessible site is improved, so that a place that previously did not provide significant experience of nature comes to do so;
- new access points are provided to sites that already provide a significant experience of nature;
- improvements are made to walking access through the areas surrounding a site, bringing more parts of developed London into the one kilometre walking distance.

As a result of the statutory plan, Greater London has the most rigorous nature conservation policies of any town or city in the United Kingdom. Now that all local authorities have a statutory duty to 'have regard' to nature conservation under the Natural Environment and Rural Communities Act 2006, the approach adopted in London provides a framework that could be applied directly to other city regions, or on a smaller scale in other towns and cities as an integral part of their sustainable development. The advantages are that it provides protection to sites of regional and local importance, recognizes the importance of green corridors throughout the urban environment, and provides a mechanism for dealing with areas of deficiency of wildlife.

The London Biodiversity Action Plan

The London Biodiversity Action Plan was launched in 1996 and has been developed in parallel with, and complementary to, the work of the London strategy. Led initially by the London Ecology Unit, and since 2000 by the Greater London Authority, it has developed programmes for the conservation of priority habitats and species that are of particular significance for London. The action plan is developed and implemented by the London Biodiversity Partnership, consisting of a wide range of bodies with an interest in biodiversity.

Members of the London Biodiversity Partnership:

Greater London Authority
British Trust for Conservation Volunteers
English Nature
London Wildlife Trust
Royal Parks Agency
Wildfowl and Wetlands Trust
London First
London Boroughs
British Waterways
Environment Agency
London Boroughs Biodiversity Partnership
Association of London Government
Community Initiatives Partnership
London Natural History Society
Thames Water Utilities Ltd
Royal Society for the Protection of Birds
London Underground Ltd
Thames Estuary Partnership
Peabody Trust

In January 2000 the Partnership published volume 1 of the action plan, which provided an audit of the key habitats and species in London. In 2001 this was followed by volume 2 (London Biodiversity Action Plan (2001)) providing detailed Action Plans for Priority Habitats and Species. Both these documents can be found on the Partnership's website (www.lbp.org.uk).

The London Biodiversity Action Plan:

Habitat Action Plans	*Species Action Plans*
Woodland	Bats
Chalk Grassland	Water Vole
Heathland	Grey Heron
Wasteland	Peregrine
Acid Grassland	Sand Martin
Tidal Thames	Black Redstart
Canals	House Sparrow
Churchyards and Cemeteries	Stag Beetle
Private Gardens	Tower Mustard
Parks, Squares and Amenity Grassland	Mistletoe
	Reptiles
	Black Poplar

The selection of priority habitats for action is based broadly on the following criteria:

- provides a good opportunity for Londoners to enjoy contact with nature (e.g. Private Gardens; Parks and city squares; Churchyards and Cemeteries);
- example of UK priority habitat within London (e.g. Heathland and Acid Grassland);
- supports a diversity of fauna and flora, including uncommon species (e.g. Chalk Grassland);
- is under threat in London (e.g. Heathland);
- presents opportunities for habitat restoration and enhancement (e.g. Heathland).

Criteria for the selection of priority species include the following:

- culturally valued and appealing, offering an opportunity to raise greater public awareness of biodiversity (e.g. Grey Heron, Mistletoe);
- characteristic of London (e.g. Black Redstart);
- priority species in UK BAP with a significant population in London (e.g. Tower Mustard, Black Poplar, Bats, Water Vole and Stag Beetle);
- substantial recent decline (e.g. House Sparrow);
- restricted distribution in London (e.g. Reptiles);
- easy to monitor (e.g. Sand Martin).

In addition to these specific action plans, the partnership has identified a number of generic actions, including site management, habitat and species protection, ecological monitoring, biological recording and communications, for each of which specific programmes have been developed.

The overall programme provides a set of targets for urban nature conservation in the capital aimed at conserving and enhancing the status of critical habitats and species, and encouraging greater public participation in wildlife conservation. Some of the action plans may seem very specific, but are actually addressing broad issues of urban development. The plan for black redstarts is a case in point. These nationally protected birds tend to be associated with derelict buildings and urban wasteland. With the pace of urban redevelopment increasing, their habitat is under threat. However, one of the options to cater for this species is to create artificial wasteland habitats on roofs of buildings. Such 'black redstart roofs' or 'brown roofs' are now being developed locally in the Thames Gateway to cater for this species, and with it no doubt a host of other wasteland specialities (Frith and Gedge 2000). The conservation of wasteland habitats could well be facilitated more widely through the use of such roofs, and also through the careful design of new industrial units to accommodate a range of other artificial habitats (Johnston and Newton 1993). Such proposals are advocated by the London Development Agency (2004) in its guidance on biodiversity for developers. The environmental benefits of 'green roofs' goes well beyond their importance for biodiversity since they have clear benefits for flood alleviation and water conservation, and also for ameliorating the urban microclimate and reducing the heat island effect in cities (Ecoschemes 2003).

Conclusion

The London Strategy and the Biodiversity Action Plan provide a powerful combination that ensures that biodiversity conservation is addressed effectively in the capital. The strategy ensures that proper consideration is given to biodiversity at the local level in development planning, while the action plan sets targets for conservation of critical species and habitats. London is fortunate in having a legal requirement for a biodiversity strategy as an integral part of its statutory plan. The model described could be applied to good effect in city regions elsewhere in the United Kingdom. It not only provides the basis for protection of important sites but also identifies wildlife corridors, many of which are important in terms of their ecological functions for flood alleviation and climatic amelioration. Given the projected impacts of climate change, the conservation of such corridors and networks is likely to become increasingly important in cities throughout the United Kingdom (Hulme *et al.* 2002). Recognition of the potential contribution of wildlife habitats to combating the adverse effects of climate change in urban areas may give renewed strength to the arguments for conservation of biodiversity. It is nearly ten years since Barker (1997) proposed green networks around towns and cities as a 'framework for the future'. The future is here and now.

References

Barker, G. (1997) *A Framework for the Future: Green Networks with Multiple Uses in and around Towns and Cities*, English Nature Research Report 256, Peterborough: English Nature.

Box, J. and Harrison, C. (1993) 'Natural spaces in urban places', *Town and Country Planning*, vol. 62: 231–5.

Department for Environment, Food and Rural Affairs (2006) *Local Sites: guidance on their identification, selection and management*, London: DEFRA.

Department of the Environment (1995) *Biodiversity: The UK Steering Group Report*, vol. 1, *Meeting the Rio Challenge*, London: Her Majesty's Stationery Office.

Ecoschemes (2003) *Green Roofs: Their Existing Status and Potential for Conserving Biodiversity in Urban Areas*, English Nature Research Reports 498, Peterborough: English Nature.

Frith, M. and Gedge, D. (2000) 'The black redstart in urban Britain: a conservation conundrum?', *British Wildlife*, vol. 11, 6.

Goode, D.A. (1989) 'Urban nature conservation in Britain', *Journal of Applied Ecology*, vol. 26, 859–73.

Goode, D.A. (1999) 'Habitat survey and evaluation for nature conservation in London', *Deinsea*, vol. 5, 27–40 (Natural History Museum of Rotterdam).

Goode, D. (2005) 'Connecting with nature in a capital city: the London Biodiversity Strategy', in T. Trzyna (ed.) *The Urban Imperative*, Sacramento: California Institute of Public Affairs.

Greater London Council (1985) *Nature Conservation Guidelines for London*, Ecology Handbook 3, London: Greater London Council.

Handley, J., Pauleit, S., Slinn, P., Baker, M., Jones, C., Lindley, S. and Barber, A. (2003) *Accessible Natural Green Space Standards in Towns and Cities: A Review and Toolkit for Their Implementation*, English Nature Research Reports 526, Peterborough: English Nature.

Harrison, C., Burgess, J. Milward, A. and Dawe, G. (1995) *Accessible Natural Greenspace in Towns and Cities: A Review of Appropriate Size and Distance Criteria*, English Nature Research Reports. 153, Peterborough: English Nature.

Hulme, M., Jenkins, G.J., Lu, X., Turnpenny, J.R., Mitchell, T.D., Jones, R.G., Lowe, J., Murphy, J.M., Hassell, D., Boorman, P., McDonald, R. and Hill, S. (2002) *Climate Change Scenarios for the UK: The UKCIP02 Scientific Report*, Norwich: Tyndall Centre for Climate Change Research, University of East Anglia.

Johnston, J. and Newton, J. (1993) *Building Green: A Guide to Using Plants on Roofs, Walls and Pavements*, London: London Ecology Unit. (Reprinted in 2004 by the Greater London Authority.)

Local Issues Advisory Group (1997) *Guidance for Local Biodiversity Action Plans: An Introduction*, Guidance Note 1, London: Local Issues Advisory Group of the UK Biodiversity Group.

London Biodiversity Action Plan (2000) *The London Biodiversity Audit*, vol. 1, *London Biodiversity Action Plan*, London: London Biodiversity Partnership.

London Biodiversity Action Plan (2001) *The Action*, vol. 2, *London Biodiversity Action Plan*, London: London Biodiversity Partnership.

London Development Agency (2004) *Design for Biodiversity: A Guidance Document for Development in London*, London: London Development Agency.

Mayor of London (2002) *Connecting with London's Nature: The Mayor's Biodiversity Strategy*, London: Greater London Authority.

Mayor of London (2003) *Green Capital: The Mayor's State of the Environment Report for London*, London: Greater London Authority.

Mayor of London (2004) *The London Plan: Spatial Development Strategy for Greater London*, London: Greater London Authority.

Office of the Deputy Prime Minister (2002) *Planning Policy Guidance 17: Open Space, Sport and Recreation*, London: Office of the Deputy Prime Minister.

Teagle, W.G. (1978) *The Endless Village: The Wildlife of Birmingham, Dudley, Sandwell, Walsall and Wolverhampton*, Shrewsbury: Nature Conservancy Council.

Town and Country Planning Association (2004) *Biodiversity by Design: A Guide for Sustainable Communities*, London: Town and Country Planning Association.

Tucker, G., Ash, H. and Plant, C. (2005) *Review of the Coverage of Urban Habitats and Species within the UK Biodiversity Action Plan*, English Nature Research Report no. 651, Peterborough: English Nature.

8 Engaging publics in environmental planning

Reflections on geographical research in a period of policy flux

Carolyn Harrison and Jacquie Burgess

Introduction

The promotion of public participation is one of three broad ideologies in British planning, alongside the protection of property rights and the advancement of public interest. Historically, conflict between these competing ideologies has left public participation the loser, although successive governments have attempted to provide more opportunities for formal and informal public involvement in the planning process. The Thatcherite years saw calls for increased participation underpinned by an ideology of consumerism and increased efficiency in service delivery, coupled with a desire to limit the constraints that planning placed on market-driven activities. Since the election of a Labour government in 1997, successive Blair governments have linked greater participation in public affairs with a devolutionary agenda and a desire to achieve greater accountability in the conduct of public services. At the same time, however, they have continued to laud the benefits of public–private partnerships, championed by previous Thatcherite governments in recognition that the capacity to get things done does not rest solely on the power of government authority. One consequence of this now well-established move from government to governance has been the blurring of the boundary between the public and the private, as non-state actors become intimately involved inside the policy process, including the process of land use and spatial planning (Stoker 1999a, b; Murdoch and Abram 2002). In theory, this shift to working in partnerships offers more opportunities for different organisations and members of the public to participate actively in the planning process. However, it is unclear how power and influence are exercised in these arrangements and whether ordinary members of the public can influence decision-making.

This dynamic policy context forms the broad backdrop to our chapter. By drawing on the findings of several research projects undertaken throughout the 1980s and 1990s, it explores ways of bringing voices routinely ignored in environmental decision-making into planning processes. Our specific focus is on public engagement with environmental planning, for over this same period growing public awareness of environmental issues, the increasing professionalism of environmental non-governmental organisations (ENGOs) and the requirement that central government and national agencies respond to European directives on the environment gave added weight to environmental matters in the planning process. By the end of the 1990s there was growing government support for sustainable development that sought to integrate environmental, economic and social policy domains, and an increased expectation that public participation would inform these policies (Her Majesty's Government 1999). As a result, environmental agencies such as the Countryside Agency (formerly the Countryside Commission) and English Nature (EN) (formerly the Nature Conservancy Council) found themselves faced not only with new environmental responsibilities and a changing political context within which to achieve them, but also with mounting public pressure to put in place policies and practices to address the continuing loss of wildlife, damaging impacts of pollution and global environmental change.

Trajectories in policy development that call for greater public accountability require a debate about 'who' participates and should participate in policy processes, but they also raise questions about the conduct and legitimacy of participatory processes of public engagement. Unless they address these issues, public agencies are confronted with the risk that expertly informed, top-down plans to advance environmental initiatives would be interpreted as bureaucratic or professional self-interest, thus further alienating public opinion. Hence, working with other stakeholders and members of the public in policy-making processes to ensure fair and transparent outcomes requires these agencies to engage more directly with the public than was the case in the 1970s, when planning and policy development were largely regarded as technical processes driven by experts and professionals (Rydin 2003). One of the purposes of this chapter is to rehearse how, through a number of projects undertaken with national agencies, our research has pointed to ways in which policies for the environment can be better informed by engaging with the concerns and understandings of different publics.

A second objective is to explore what members of the public understand they achieve by taking an active part in planning processes. In practice, the point at which most people feel moved to participate is at the sharp end of local planning practice, namely development control. At this level, public participation in land-use planning received a boost in the Town and Country Planning Act 1990, with its move to plan-led development, and in the Planning and Compensation Act 1991. This latter Act introduced a statutory requirement for all types of planning application to be publicised and for there to be opportunities for members of the public to make their views known by writing to the relevant local authority (LA). Public consultation is accomplished by written objection and although as part of its modernising agenda the Labour government has encouraged LAs to provide electronic-based services that allow members of the public to register objections remotely, including by email, public opposition is expected to be voiced indirectly rather than by face-to-face discussion (Department of Environment, Transport and the Regions 1998a, b, c). Recent revisions to the planning system replaced local plans with Local Development Frameworks (LDFs), with the requirement that a statement of the community involvement process must accompany LDFs (Department of Transport, London and the Regions 2001). Closely linked with the activities of Local Strategic Partnerships, community involvement, including in the preparation of open space strategies for example, will be enmeshed in a local politics that reflects central government's commitment to new private–public arrangements.

Running the risk that these new procedures will strengthen the presence of 'the usual suspects', rather than enlisting new participants, these revisions to the planning system have also been accompanied by other initiatives seeking to streamline and speed up the planning process (Davies 2001). For example, Best Value policies seek to achieve efficiency gains in public service delivery, including the planning system, and require LAs to consult regularly with the public (Stoker 1999b). Although written questionnaires sent to households remain the conventional method of surveying public opinion, there is evidence that LAs have experimented with more innovative ways of engaging the public, for example through citizens' panels, citizens' juries, focus groups and community visioning exercises (Lowndes *et al.* 2001a, b). Nevertheless, it is not clear to what extent local decisions have been influenced by these novel participation initiatives or whether local people feel empowered by their experience. By working with people who had already taken an active part in planning processes, our research sought to explore the relationships between new procedures for involving the public and an understanding of what participation achieved.

Our final objective is to introduce a critical dimension to thinking about how the environment and nature are valued in the planning process. This is a complex question, for we understand values as the reasons given for justifying proper conduct in relation to the living world *and* as their embodiment in social actions and organisations that support these values. Understood in this way, approaches to valuing nature involve not just the basis for determining or characterising what counts as nature, but the methods used to express nature's value and the preferred means of protecting it. When values become institutionalised, attempts by different agencies and organisations to engage the public in environmental decision-making are likely to be predicated on the preconditions of their own approaches to valuing nature. In turn, different publics find their own approaches to valuing nature foreclosed by the methods and procedures commonly used to extend public participation in decision-making. As a result, public voices may be sought but unheard.

In taking forward this intellectual strand in our research, we first discuss a framework for thinking about how nature is valued using the threefold typology illustrated in Table 8.1. This typology, published in 1999, captured the experiences that we, together with our colleague Judy Clark, had gained through a series of empirical research projects addressing questions of how expert and lay knowledges, meanings and values for nature were communicated in different decision-making forums and contexts (Harrison *et al.* 1999). The typology gave some much-needed clarity in analysing situations where participants seemed constantly to be 'talking past' one another.

Table 8.1 Typology of approaches to valuing nature

Value of nature determined by:	Value of nature expressed through:	Nature best protected by:
Science	Scientific criteria	Standards and regulations (legislation)
Individual use (preferences)	Willingness to pay/accept (contingent valuation)	Economic efficiency (cost–benefit analysis)
Common good, ethical, moral concerns	Social consensus	Political process

Source: Harrison *et al.* (1999).

Valuing nature

Scientific approach

The dominant approach to valuing nature in Britain is based in science. As espoused through the policies and practices of organisations such as the Nature Conservancy Council (NCC) and its successor body English Nature (EN), and by ENGOs such as the Royal Society for the Protection of Birds (RSPB), valuing nature depends on the use of scientific criteria to identify 'valued nature' and the use of legislation to protect key sites: Sites of Special Scientific Interest (SSSIs), Special Areas of Conservation (SAC) and the like. Derek Ratcliffe's monumental *Nature Conservation Review* (1977) provides the reasoned rationale for this approach, in which objects (species) and their attributes (rarity, diversity) are used to determine 'key sites' that require legal protection. The approach is based on systematic field survey and inventory, and receipt of expert appraisal by natural scientists; it achieves legitimacy through the authority that policy-makers and politicians accord to science. In the sense that all judgements based in science are predicated on the provisionality to which science admits, not all conservation scientists agree with the key site approach – witness, for example, growing support for a 'wilding approach' to valuing nature (Adams 2003). 'Wilding' places value on the inherent (wild) trajectories of biological change possessed by ecosystems and extensive tracts of land, rather than on rare and declining species associated with particular sites. But whichever approach is promoted, by laying claim to being an objective, rigorous approach to valuing nature, a scientific approach makes claims to universal values that society as a whole may not share. For example, through our work on the conflict between nature conservationists and developers over the future of Rainham Marshes SSSI, and our subsequent work on a Wildlife Enhancement Scheme on the Pevensey Levels in Sussex, we discovered that the scientific discourse for valuing nature was not universally regarded as legitimate by all actors (Harrison and Burgess 1994; Harrison *et al.* 1998; Burgess *et al.* 2000).

The bruising encounters experienced by EN's scientists in the quasi-legal arenas of public planning inquiries are testimony to the double-edged sword that reliance on science becomes when it moves out of the laboratory and into the social and political arena. Unless accompanied by the political will to support science and the resources to implement policies, this approach to valuing nature is open to challenge. During the 1980s the scientific approach to valuing nature came under increasing attack in the United Kingdom, most prominently in 1986, when Nicholas Ridley as Secretary of State threatened to privatise NCC (Marren 2002). Regarded by the government of the day and the private sector as an obstacle to development, SSSIs became a political issue. At the same time, the proposition that the costs of protecting nature should be fully accounted for in decision-making gained ground. This rationale underpins a second approach to valuing nature.

Individual use approach

The individual use approach determines the value of nature through the sum of individual preferences. Closely linked with the emergence of environmental economics as a sub-discipline and with the methodologies of cost–benefit analysis (CBA), this approach is consistent with achieving economic efficiency in decision-making. Much favoured by the former Department of Transport for assessing the costs and benefits of road schemes, it is also associated with concepts of 'no net loss' and 'mitigation' incorporated into the European Habitat Directive of 1994, and with the payments to farmers made through various agri-environment schemes promoted by central government and the European Union.

CBA is an attractive approach because it has the apparent ability to express all costs and benefits of a development in the common unit of money. Moreover, through the use of contingent valuation methods (CVM) as a means of determining the monetary value of nature, members of the public are invited to indicate what they would be willing to accept by way of compensation were an area to be damaged or lost. In this way the method claims to be democratic, but has been widely criticised by institutional economists such as Jacobs (1994) and geographers such as Adams (1995). The basis of much of the criticism stems from the assumption that the multiple values nature has for society can be captured in a single monetary equivalent. At the same time, CV methods rely on averaging the sum of individual preferences to represent how society as a whole values nature. Privileging individual preferences rather than inviting consideration of collective responsibilities, the approach supports decisions based on economic efficiency rather than questioning the correctness of the outcome CBA suggests. In a project supported by ESRC and English Nature, critiquing a CV survey of the Pevensey Levels Wildlife Enhancement Scheme, our research provided strong evidence to suggest that CVM was not the 'democratic' and 'participatory' mechanism implied by the environmental economists. For many participants in our study, the monetary 'value' they produced for the survey was not meaningful but rather an artefact of the method (Burgess *et al.* 1998, 2000; Clark *et al.* 2000).

Common good approach

A third approach has its basis in understanding nature as a common good, and requires social consensus to be reached about how the environment and nature are to be valued. Thereby, it is an approach dependent on a political process for articulating and supporting consensus. In comparison with the two other approaches, more emphasis is given to moral and ethical grounds for valuing nature and to deliberative methods of reaching agreement about what actions can be supported towards the environment. With its emphasis on deliberation, the approach fits most closely with forms of discursive democracy (Dryzek 1990; Munton 2003) and with collaborative planning advocated by Forester (1999) and Healey (1997). First developed in 1997 by the four national agencies – the Countryside Commission, English Nature, the Environment Agency and English Heritage – and variously called Environmental Capital, Quality of Life Capital and Quality of Life Assessment (QOLA), QOLA is an example of a common good approach to valuing the tangible and intangible properties of nature, places and landscapes and of achieving consensus about the collective values society assigns to them (Levett 2006). In practice, too, some of the more innovative experiments in public participation, such as citizens' juries (Coote and Lenaghan 1997), consensus conferences (Joss and Durrant 1995) and deliberative mapping (Burgess *et al.* 2004), have their origin in this approach.

While it represents a simplification of a complex set of interrelationships, this framework helps to reveal how processes and procedures of valuing nature and the way society is organised are not independent of each other; rather, they are constituent of each other, often in ways that are hidden from view and taken for granted. Of the three approaches, the scientific and individual preference approaches have dominated public decision-making, not least through their incorporation into the standard procedures and practices adopted by central and local government. By comparison, common good approaches, though they often resonate with public concerns about nature, are poorly represented in environmental decision-making. In acknowledging the legitimacy of each of these approaches, but seeking to give a more effective voice to common good approaches, the challenge for us as researchers has been twofold. First, we needed to develop ways of engaging members of the public in meaningful discussions about their own approaches to valuing nature. Second, we sought to establish mutually beneficial relationships with policy-makers and environmental organisations, so that our research might gain some purchase on policy development and environmental decision-making. In the following sections we tell the story of how, through a succession of research projects, we took forward this research agenda.

Enjoying the countryside

Our preferred approach to engaging members of the public has been pursued through providing discursive spaces in which people have an opportunity to talk about and discuss their environmental attitudes, meanings and values in some depth, and in ways that provide participants with an opportunity to learn from this experience. Consistent with the deliberative and consensual dimensions of a common good approach to valuing nature, we developed the use of in-depth groups based on the principles and practices of group analysis, and then developed by Abercrombie (1983) for use in 'task-oriented' small groups (Burgess *et al.* 1988a, b). An in-depth group provides opportunity for people to talk about their experiences in a supportive environment, and over several meetings group members have time to explore, reflect on and sometimes revise their reasons for valuing nature, places and landscapes. Personal, social and cultural meanings of open space, parks and countryside for different groups of people in different localities are expressed in ordinary conversation. Transcripts of the discussions form the basis of a textual analysis that is guided by the grounded theory of Glaser and Strauss (1967). Analysis provides a structured interpretation of shared meanings and values, together with their underlying rationales and signification.

We pioneered the use of in-depth groups in a research project in 1985–7 designed to develop a methodology for evaluating amenity land in the urban fringe funded by the (then) Social Science Research Council (SSRC) together with the (then) Countryside Commission. At this time the Commission was beginning to build a reputation for innovative social research to support its work on recreation and access (Phillips and Ashcroft 1987). Our research aimed to work with groups of residents in the London Borough of Greenwich. It was designed, first, to explore the meanings and values that residents attached to the different kinds of green and open spaces in their local environments, and their use of the wider countryside. The study was to inform a major review of national recreation policies published in 1987 (Countryside Commission 1987). At this early stage of the introduction of qualitative research evidence into the planning process, the Commission employed a market research consultancy to undertake a focus group study of attitudes to the countryside. The in-depth groups we proposed to conduct in Greenwich with four groups of eight to ten people would meet for six weeks and offered a new approach to researching such questions.

As part of our contract with participants in the Greenwich study we wrote up the main themes discussed in each group and sent them to participants. These group reports were also sent to Commission officers. Written in an accessible style with quotations and extended extracts from the discussions used to illustrate key themes, the reports were welcomed by officers as an authentic public voice, 'a breath of fresh air' and 'just the kind of findings previous quantitative surveys had failed to reveal'. But mindful of their own policy requirements, officers were also keen to circumnavigate the theoretical implications of the socially and culturally laden interpretations of the different natures and countrysides that our work highlighted. What they wanted was 'a pale shadow for policy' (Burgess *et al.* 1990). In practice the Commission accepted an early paper published in *Landscape Research* (Harrison *et al.* 1986) as our main input to the policy review process.

As revealed through our empirical study, the concept of countryside and open space as both imaginative and literal *gateways to a better world*, a world that compares favourably with the burden of urban life, was the single most pervasive meaning that resonated most closely with the values and experiences expressed by all group members. The relative emphasis people placed on what constituted 'a better world' was affected by gender, personal experiences and economic and social factors, but poorly explained by social analyses that focused on class. In practice, too, the strength of signification given to the social function of open spaces and countrysides was unexpected. For example, while conventional representations of enjoying the countryside as a solitary activity were recognised by some group members, many members gained more enjoyment in the company of sympathetic others, family and friends. Often, sharing enjoyable experiences with others made countryside visits all the more memorable.

Taken together, the cultural and social values group members attached to different kinds of countryside offered strong public support for policies that would ensure the farmed and settled countryside, not just the fine landscapes of open countryside, would be planned and managed in ways that allow people to share in the 'better world' they represent. Coming at a time when agricultural intensification was inflicting heavy losses on wildlife habitats and contributing to widespread landscape change, revealing the significance and value these same landscapes had for public enjoyment provided a legitimate platform on which to build policies for enjoying the countryside (Countryside Commission 1987). Acknowledged by members of the policy review team as having made a difference to policy development, this impact occurred with an immediacy that most policy-relevant research seldom achieves (Phillips and Ashcroft 1987; Owens 2005).

With the benefit of hindsight, we recognise that without the willingness of the Commission to run the risk of supporting innovative social research and the window of opportunity provided by the review of its recreation policies, this outcome might not have been the case. The authenticity of the public voices that Commission officers acknowledged accrued from its basis in a common-good approach to valuing nature, especially through the deliberations that working with in-depth groups permits. Nevertheless, the 'popular values' for nature, countryside and open space rehearsed in written papers exist as our interpretations that distance group participants from direct engagement with the policy process. In numerous presentations to Commission staff and conferences, our role became one of 'standing in' for the public, a position with which we became progressively more uncomfortable over our research career (see, for example, Davies and Burgess 2004).

A second, and unanticipated, opportunity to contribute to policy development that stemmed from the Greenwich study, but with very different outcomes, came in 1994 with a contract from English Nature to review size and distance criteria for accessible green space in cities.

Accessible green space

The activities of NCC and English Nature in urban areas have always been limited because successive administrations have given precedence to country-side issues and statutory (re)notification of key sites; the overwhelming majority of SSSIs are in countryside locations. Nevertheless, in his capacity as EN's Urban Adviser, George Barker succeeded in advancing the cause of urban wildlife within the organisation throughout the 1980s and 1990s. In the process he did much to expose the social and cultural values underpinning objective, 'scientific' approaches to valuing nature, especially attitudes to introduced and alien species that occur spontaneously in 'natural' communities throughout urban areas. Treading a fine line between activist and expert adviser, George Barker recognised that the scientific culture of an organisation like English Nature would demand evidence-based research to support the development of novel urban policies. English Nature's contract to review size and distance criteria for accessible green space in urban areas was designed to do just that.

The contract followed the publication of a paper by Box and Harrison (1993) that set out the case for a spatial standards approach to urban green space planning, prepared in part through work undertaken for the UK Man and the Biosphere Urban Forum (2006). This paper drew on John Box's professional experience of local nature reserves in the West Midlands and our own research on urban green spaces in Greenwich (Burgess *et al.* 1988c; Harrison *et al.* 1987). Although urban land-use planners were familiar with a spatial hierarchy approach to formal open space provision, few approaches in the urban area addressed the contribution that more naturalistic settings and places make to people's enjoyment. This is despite the fact that guidance issued by English Nature on local nature reserves (English Nature 1991–5) suggested they could be established if the wildlife or natural features of a site were special by virtue of the fact the people found them so for the quiet enjoyment and appreciation of nature. Consistent with the common good values espoused in this latter approach, the standards framework proposed by Box and Harrison (1993) advocated a needs-based approach, one that reflected the attitudes and values of local people.

Carried out with the help of other researchers – Gerald Dawe and Alison Millward – our contract research reviewed both natural and social scientific research to support benchmarks of provision. By combining both evidence from natural and social sciences the review established that physical access to natural space involved much more than distance alone (Harrison *et al.* 1995). The Greenwich study had shown that the supply of open space in a neighbourhood influenced people's use of parks. It also showed that whether an open space was regarded as accessible or not depended on a number of constraining factors. Accessibility depended both on users' physical ability to gain access to sites, which often involved negotiating circuitous routes with poorly located crossing and access points, and on a suite of social and cultural factors. In the case of children, for example, the distance from home that unaccompanied children were permitted to roam by their parents meant that accessible spaces were limited to those spaces 'on the doorstep'. A more complex set of constraints served to deter other groups of adults from visiting even nearby spaces. In particular, the research pointed to the behavioural consequences of perceptions of open spaces as 'risky places' by some groups in society: women, elderly people and members of ethnic minorities.

A common paradox emerged through the groups in our Greenwich study and through other groups conducted in research for the Community Forest Unit (Burgess 1995). Although making use of naturalistic local spaces was recognised as beneficial by people who currently do not use these spaces, factors serving to limit access affected different groups in different ways. In other words, while public policy addressed a homogeneous public, as revealed through our research, 'access' to public open space is as much about social exclusion as about physical distance. Taking this approach forward into a standards approach meant that the maximum distance unaccompanied children were permitted to walk to gain access to a natural open space (300 metres) formed the benchmark for local provision. The standard was extended to encompass a spatial hierarchy of sites based on studies of the reported distances used by accompanied children and adults. The suggested minimum standard is illustrated in Table 8.2.

English Nature did not actively promote these recommendations, in the main because in the wake of the Wildlife and Countryside Act 1991 its priorities continued to lie with the work of renotifying all existing and proposed SSSIs (Marren 2002). Under these circumstances the research report was 'put on the shelf', lacking a champion to take the standards approach further forward. This was despite the research suggesting a firm foundation for a standards approach, with evidence for distance criteria being well supported by research in the United Kingdom and elsewhere in Europe. But something in the political mood changed in the early years of the new century to give our 'old' research new saliency.

The government's long-awaited Urban White Paper was published in 2000 and was followed by a number of other policy initiatives (Department of Environment, Transport and the Regions 2000). The Urban Green Spaces Task Force was set up to advise on proposals for improving the quality of urban parks, play areas and green spaces (Urban Green Spaces Task Force 2002), and

Table 8.2 Minimum standards for accessible green space

An accessible natural green space less than 300 m (in a straight line) from home

An accessible 20-ha site within 2 km of home

An accessible 100-ha site within 5 km of home

An accessible 500-ha site within 10 km of home

At least 1 ha of local nature reserve for every 1,000 people

Source: Harrison *et al.* (1995).

a flurry of activity centred in the Office of the Deputy Prime Minister linked green spaces, social inclusion and sustainable communities (Frith 2003). With government's espoused determination to take forward a more inclusive agenda to policy development, and to promote the use of indicators as a means of tracking progress towards service delivery and public performance, new 'spaces' for the development of socially inclusive environmental policies such as 'accessible nature' seemed to open up.

English Nature responded to this shift in government policy and in 2001 commissioned a review of the earlier research on accessible natural green space standards. Undertaken by the Centre for Urban and Regional Ecology at Manchester, this new research revalidated our earlier findings, and the structure of the standard itself withstood scrutiny (Handley *et al.* 2003). Now set within a rapidly changing political and policy context that required local authorities to develop means of measuring service delivery, the new report developed a tool kit for assessing and developing a model for local accessible natural green space. Official endorsement of the tool kit came with its inclusion in *Assessing Needs and Opportunities: A Companion Guide to PPG17* (Office of the Deputy Prime Minister 2002a), published alongside *Planning Policy Guidance 17: Open Space, Sport and Recreation* (Office of the Deputy Prime Minister 2002b). Among other approaches listed in the guide, English Nature's model is the only one with an explicit focus on the natural environment. For a national agency that had the reputation among metropolitan authorities and urban wildlife trusts of 'failing to lead' on matters of urban wildlife (Marren 2002), this is quite an achievement. Time will tell whether the priorities of Natural England, the new agency formed by the merger of the Countryside Commission and English Nature in 2006, will continue to lie in the urban area.

As researchers, we came to question why research on valuing accessible green space as an issue of social exclusion evolved into a 'tool kit' for assessing needs and provision. For an organisation like EN, whose institutional values lie in science, a standards-setting approach meshes with an appeal to the universal values to which science makes claims, and a tool kit offers a technical method of applying objective criteria that appeals to experts and professionals. If used sensitively, the tool kit could make a difference to the quality of people's lives. But in a world where local authorities are faced with hundreds of targets to meet as part of Best Value performance, efficiency and not sensitivity to the task in hand is likely to drive the process of assessment. So, while the government's signalled intention to develop a new indicator for 'access to local green space' (National Audit Office 2006) can be welcomed, expectations that this indicator will acknowledge the social and cultural determinants of *accessible* natural green space are unlikely to be met.

Reflections

These examples of projects taking forward research with in-depth groups into environmental policy arenas position researchers as the conduit through which public values are conveyed to policy-makers. Standing proxy for different publics in this way always runs that risk that knowledge will be used strategically by policy-makers to suit their own purposes (In't Veld 2000). However, even in situations where policy-makers are committed to full public participation, a growing body of research demonstrates how environmental policy-making processes are permeated with issues of systemic power that position researchers and the public on the margins of the decision-making process (Flyvbjerg 1998; Hillier 2000). This is all the more so at the level of local government, where a process of 'hollowing out' of the state and the formation of public–private development partnerships have generated governance arrangements that blur boundaries between the public and the private, and between those interests being pursued 'within' the policy process and those 'external' to it (Rydin *et al.* 2003). In this new policy context we were keen to understand what members of the public who had taken an active part in planning processes believed their actions achieved.

Public participation and the politics of an urban renaissance

Lord Rogers' 1999 Urban Task Force report *Towards an Urban Renaissance* and the government's Urban White Paper published the following year (Urban Task Force 1999; Department of Environment, Transport and the Regions 2000) reaffirmed the strength of economic arguments for regenerating cities. Their advocacy of higher building densities and the use of brownfield sites as the location for future development prompted strong opposition from a number of environmental agencies and organisations. Based on economic efficiency, the government's valuation of brownfield sites as wasting assets that development can bring into beneficial use discounts a more pluralistic understanding of such sites as communal resources in which all sections of society have an interest. For many nature conservationists in the voluntary sector, brownfield sites represent the few examples of truly wild and spontaneous natural communities to be found in urban areas. Basing their arguments in common good approaches, they argue that as 'urban commons', brownfield sites are as much part of our natural and cultural heritage as many other 'artificial' habitats in the United Kingdom (Gilbert 1991).

In the context of cities like London, however, the ascendancy and authority of the economic argument for development, coupled with new governance arrangements that place some interests inside the development process and others outside, mean that it is seemingly impossible for other ways of valuing nature to gain purchase on the development process. Our research conducted with a network of environmental groups in London in 1999 and 2000 suggests that there is little agreement among organisations outside the development process about how to engage with these new development threats and governance arrangements (Harrison and Davies 2002). Some organisations looked to their good relations with the LA to ensure that policies and best practices in consultation were followed; others preferred a more adversarial stance, since in contexts where 'planning gain' arose, the LA was both poacher and gamekeeper. Yet other members sought opportunities to work with the developer to secure the best mitigation package possible. Requiring a sophisticated and dedicated approach to working with complex procedures and in a time-frame that prevented detailed environmental appraisal, members of the network understood that the changing policy context of development would mean inevitable losses of biodiversity.

Revisions to the Planning System also had implications for public participation (Department of Transport, London and the Regions 2001, 2002). Developers were expected to take more responsibility for consulting on proposals, and LAs were expected to provide new information and communication technologies (ICTs) that favour remote rather than face-to-face communication. Our research suggests that the effectiveness of these procedures for improving public participation is highly debatable (Bedford *et al.* 2002; Harrison and Haklay 2002).

The research project tracked two case studies through the development process to the point of decision. In both cases, applications were for residential development on brownfield sites and received approval from the relevant LA. One small site involved building on land that was part of a local nature reserve; a second, larger site alongside the Thames involved building on land identified by the developer's ecologist as having 'more than local importance for nature conservation', although not recognised as such in the local plan. Both applications contravened a number of local policies in terms of building densities, elevations and mix of uses, and environmental measures negotiated as part of each application were material considerations to the decision process.

Consistent with the procedures recommended by government for raising public participation, developers convened public meetings at which their proposals were presented and discussed. For the larger, riverside development, a more extensive approach to public consultation was involved. Consultants for the developer held focus groups with local people, individual households within a 400-metre radius of the site were sent a questionnaire, and over 50 meetings were held with amenity and environmental groups. Our research involved in-depth interviews with members of the LAs and the developers' teams, with agencies and organisations that had objected to the proposals, and members of the public. In addition, with support from one LA, we also convened workshops with members of the public to explore their attitudes to the use of a purpose-built ICT system as a 'participatory tool' in the planning process.

Public participation and developer-led consultation

Reactions of members of the public to the consultation processes they had experienced were permeated with questions about lack of transparency and fairness in the development process. For example, members of amenity organisations believed the LA to be receptive to pubic concerns only at key points in the electoral cycle, rather than to calls to increase their responsiveness to everyday public concerns. For many individual objectors too, the decisions of the council appeared to be without vision and to lack any responsiveness to local needs; instead, development seemed to be 'entirely commercially driven'. In both cases, individual objectors could not recall which party had contacted them (the developer, the council or both parties), and people who had attended the public exhibitions mounted and attended by the developers simply saw the displays as promoting 'the commercial interest ... trying to minimise criticism'. Instead, the public believed the LA and not the developers were the people to whom their objections should be rightly addressed.

Members of amenity and environmental organisations recognised that in the riverside case the public consultation undertaken by the developer had been more than usual. But they were equally sceptical about the meaningfulness of that consultation. As one member of an amenity group put it:

> I would say that the public consultation, which has been praised, has been a bit of a false exercise. Because, in real terms, what has it achieved? It's created a pleasant atmosphere. It's created a certain amount of positive response in that people are saying: 'Oh well, the scheme is much better than the one before', which is true. But is it good enough?

Many respondents believed that what they had participated in, as a form of developer-led consultation, was little more than 'a game', offering increased participation, some responsiveness to public concerns but no meaningful participation.

The local political context also had a strong bearing on the view of amenity organisations that were members of a partnership typical of those championed by central government for overseeing the regeneration of the area. Membership of the partnership provided access to informal networks through which the developer had a chance to gain intelligence about local opposition to its proposals, the LA was seen to be acting consistently with the advice of central government, and local organisations could gain early knowledge of details of the development. But membership also had political consequences, as one member revealed. A letter written by the chair of the partnership to a member whose organisation had objected to a planning application from the developer made it clear that 'it's none of your business'. Here, the micro-politics of power 'within' the partnership arrangement sought to suppress legitimate views organisations might wish to express 'outside' it, much as Hillier (2000) suggests.

It may well be that local planning authorities (LPAs) are not conscious that structures of power associated with new governance arrangements serve to frustrate members of the public who wish to take an active role in public affairs. Nonetheless, responses of members of the public we interviewed reveal that they are all too aware of the powerful influence of economics and institutionalised politics on the outcomes of public participation. With this understanding comes scepticism about what 'more' and 'novel' participation practice can achieve, especially when responsibility for public consultation is vested with the private sector and not the LA. More than this, when organisations in civil society are invited to join decision-making partnerships, the boundaries between the public and the private are not only blurred but open to manipulation in ways that are hidden from view (see also Rydin *et al.* 2003).

Public participation and information and communication technologies (ICTs)

In the second part of the research project, we brought nine members of the public (two men and seven women) together in a workshop designed to offer hands-on experience of a new ICT tool that was being developed as a planning tool by the LA. This was the second workshop we had convened with local residents; an earlier one with 14 participants (eight men and six women) had taken place in the design stages of the research project. Both workshops provided an opportunity for participants to discuss their experiences in a focus group held at the end of the practical sessions. Details of the ICT system, which included a geographic information system (GIS), and the way the workshops were conducted are described by Harrison and Haklay (2002). Here we focus on the issues raised by participants in discussion at the end of the workshops.

Three themes dominated the group discussions: *deepening and extending knowledge networks*; *getting behind the screen to data providers*; and *institutional practices and the role of trust*. In each case, discussions revealed a number of ways in which the technology is embedded in existing social relations and not separate from them. For example, participants were keen to discuss how access to this new technology might open up the planning process to greater public scrutiny. In performing this role, participants understood that much would depend on the accuracy and reliability of data displayed and the extent to which different data sources could be integrated, especially to reveal the environmental impact of a proposed development in its local context and the cumulative impacts of developments over time. They wanted to know whether this kind of information would be available to the public and, most of all, they wondered whether they could trust the LA to be the honest broker they expected it to be.

At the same time, many members of amenity societies and residents' groups believed that if the LA wanted members of the public to behave as active citizens, then only by making publicly available the same information and criteria that planners used in making their decisions could the public act in this role. They argued that without this kind of information being available, the development-control process would not be accepted as either transparent or fair. Working with a model of local government that places the public firmly outside the planning system but looking in, and with the planning process itself as a purely technical procedure, this model seemed unrealistic to other community activists. These latter respondents understood the development process to be a more political one involving interests and alliances that new governance arrangements condoned. Under these circumstances, having access to ICT was unlikely to act in ways that could open up the process to greater public scrutiny.

Curry (1998) argues that it is easy to be seduced by the 'mystique' of public participatory GIS, and indeed members of the public who took part in the workshops marvelled at the new sources of information that ICT opened up, and welcomed the tantalising views of the 'what if' scenarios that GIS project. But they were equally sceptical about their putative benefits as planning tools. In particular, they recognised that in order to realise the enormous communicative potential of ICT, its use would need to be extended into public arenas that brought together planners, politicians and members of the public. Without the kind of face-to-face discussions they had experienced in the workshops, respondents believed that new technologies would not become a learning resource that would benefit the whole community.

Conclusions

Taking this opportunity to reflect on our geographical research over the past 25 years, we continue to be encouraged by the emancipatory power of working in groups, especially for deepening an understanding of how members of the public value nature. But as we have shown, our experience also shows that public participation in environmental decision-making continues to be marginalised. In a period of policy flux that has seen, in theory at least, more opportunities for public participation in environmental planning than in the past, our work suggests that while public voices are frequently sought, those of the wider, 'unorganised' public are routinely unheard.

At the local level we attribute much of this 'deafness' to procedures and methods of public consultation that are shaped increasingly by approaches to valuing nature based on economic efficiency and by targets and indicators that seek universal solutions rather than context-sensitive ones. At the same time, new governance arrangements have blurred boundaries between the state, the private sector and the public, so that public participation is further enmeshed in a local politics that defeats meaningful outcomes and breeds scepticism and distrust of local government. For us, as researchers, it is the process of developing policies that is important, because deliberation changes the relationships between actors as they learn about the task in hand (Clark *et al.* 1998, 2001; Davies and Burgess 2004). Like Innes and Booher (2000), we believe that 'tools', such as indicators, cannot drive policy, because they are enmeshed in pre-given, often universalising approaches that do not capture the necessary subtleties of what valuing nature means to different groups in society. Without more investment in the conduct and resourcing of deliberative, discursive spaces that encourage collaborative approaches to local planning, we conclude that public voices will continue to be unheard.

At the national level the government has experimented with more deliberative approaches to environmental decision-making, for example by engaging members of the public and stakeholders in appraising the options for introducing commercial GM crops in UK agriculture, and for managing the United Kingdom's intermediate and high-level radioactive waste (Burgess and Chilvers

2006). The government also committed itself to undertake a deliberative forum on sustainable consumption to look at how government and citizens can work together in delivering the UK sustainable development strategy (Her Majesty's Government 2005). Experience to date suggests that politicians are often uncomfortable with these approaches and even more reluctant to act on their outcomes (Burgess 2005; Burgess and Chilvers 2006). It remains to be seen whether these experiments represent a genuine move towards deliberative democracy.

References

Abercrombie, J. (1983) 'The application of some principles of group analytic psychother-apy to higher education', in M. Pines (ed.) *The Evolution of Group Analysis*, London: Routledge & Kegan Paul, pp. 3–16.

Adams, J. (1995) *Risk: The Policy Implications of Risk Compensation and Plural Ratio-nalities*, London: UCL Press.

Adams, W. (2003) *Future Nature: A Vision for Conservation*, 2nd edn, London: Earth-scan.

Bedford, T., Clark, J. and Harrison, C. (2002) 'Limits to new public participation prac-tices in local land use planning', *Town Planning Review*, vol. 73, 311–31.

Box, J. and Harrison, C. (1993) 'Natural spaces in urban places', *Town and Country Planning*, vol. 62, 231–5.

Burgess, J. (1995) *Growing in Confidence*, Cheltenham: Countryside Commission CCP 457.

Burgess, J. (2005) 'Follow the argument where it leads: some personal reflections on "policy-relevant" research', *Transactions of the Institute of British Geographers*, vol. 30, 273–81.

Burgess, J. and Chilvers, J. (2006) 'Upping the ante: reflections on UK public and stake-holder engagement processes in GM and radioactive waste decision-making', *Science and Public Policy*, vol. 33, 713–28.

Burgess, J., Chilvers, J., Clark, J., Day, R., Hunt, J., King, S., Simmons, P. and Stirling, A. (2004) 'Citizens and specialists deliberate options for managing the UK's legacy intermediate and high level radio-active waste: a report of the Deliberative Mapping Trial, June–July 2004', www.corwm.org.uk; www.deliberative-mapping.org.

Burgess, J., Clark, J. and Harrison, C. (1998) 'Respondents' evaluations of a contingent valuation survey: a case study based on an economic valuation of the wildlife enhance-ment scheme, Pevensey Levels in East Sussex', *Area*, vol. 30, 19–27.

Burgess, J. Clark, C. and Harrison, C. (2000) 'Culture, communication and the informa-tion problem in contingent valuation surveys: a case study of a Wildlife Enhancement Scheme', *Environment and Planning, C*, vol. 18, 505–24.

Burgess, J., Goldsmith, B. and Harrison, C. (1990) 'Pale shadows for policy: reflections on the Greenwich open space project', *Studies in Qualitative Methodology*, vol. 2, 141–67.

Burgess, J., Limb, M. and Harrison, C. (1988a) 'Exploring environmental values through the medium of small groups. Part 1: Theory and practice', *Environment and Planning A*, vol. 20, 309–26.

Burgess, J., Limb, M. and Harrison, C. (1988b) 'Exploring environmental values through the medium of small groups. Part 2: Illustrations of a group at work', *Environment and Planning A*, vol. 20, 457–76.

Burgess, J., Limb, M. and Harrison, C. (1988c) 'People, parks and the urban green: a study of popular meanings and values for open spaces in the city', *Urban Studies*, vol. 25, 455–73.

Clark, J., Burgess, J., Dando, N., Bhattachary, D., Heppel, K., Jones, P., Murlis, J. and Wood, P. (1998) 'Prioritising the issues in Local Environment Agency Plans through consensus building with stakeholder groups', Project Record W4/W4-002/1, Bristol: Environment Agency.

Clark, J., Burgess, J. and Harrison, C. (2000) ' "I struggled with this money business": respondents' perspectives on contingent valuation', *Ecological Economics*, vol. 33, 45–62.

Clark, J., Burgess, J., Stirling, A. and Studd, K. (2001) *Local Out-Reach: The Development of Criteria for the Evaluation of Close and Responsive Relationships at the Local Level*, Research & Development Report, Bristol: Environment Agency.

Coote, A. and Lenaghan, J. (1997) *Citizens' Juries: Theory into Practice*, London: Institute for Public Policy Research.

Countryside Commission (1987) *Policies for Enjoying the Countryside*, Cheltenham: Countryside Commission CCP 234.

Curry, M.R. (1998) *Digital Places: Living with Geographic Information Technologies*, New York: Routledge.

Davies, A.R. (2001) 'Hidden or hiding? Public perceptions of participation in the planning system', *Town Planning Review*, vol. 72, 193–9.

Davies, G. and Burgess, J. (2004) 'Challenging "the view from nowhere": citizen reflections on specialist expertise in a deliberative process', *Health and Place*, vol. 10, 349–61.

Department of the Environment, Transport and the Regions (1998a) *Modernising Planning*, London: The Stationery Office.

Department of the Environment, Transport and the Regions (1998b) *Modern Local Government: Guidance on Enhancing Public Participation*, London: The Stationery Office.

Department of the Environment, Transport and the Regions (1998c) *Modern Local Government: In Touch with the People*, Cmd 4014, London: Department of the Environment, Transport and the Regions.

Department of Environment, Transport and the Regions (2000) *Our Towns and Cities: The Future: Delivering an Urban Renaissance*, London: The Stationery Office.

Department of Transport, London and the Regions (2001) *Planning: Delivering a Fundamental Change*, London: Department of Transport, London and the Regions.

Department of Transport, London and the Regions (2002) *Planning Green Paper*, London: Department of Transport, London and the Regions.

Dryzek, J. (1990) *Discursive Democracy: Politics, Policy and Political Science*, Cambridge: Cambridge University Press.

English Nature (1991–5) *Local Nature Reserves in England*, Peterborough: English Nature.

Flyvbjerg, B. (1998) *Rationality and Power*, Chicago: University of Chicago Press.

Forester, J. (1999) *The Deliberative Practitioner: Encouraging Participatory Planning Processes*, Cambridge, MA: MIT Press.

Frith, M. (2003) 'Will 17 be better than 9?' *Ecos*, vol. 24, 16–19.

Gilbert, O. (1991) *The Ecology of Urban Habitats*, London: Chapman & Hall.

Glaser, B.G. and Strauss, A. (1967) *The Discovery of Grounded Theory: Strategies for Qualitative Research*, New York: Basic Books.

Handley, J., Pauleit, S., Slinn, P., Baker, M., Jones, C., Lindley, S. and Barber, A. (2003) *Accessible Natural Green Space Standards in Towns and Cities: A Review and Tool Kit for Their Implementation*, English Nature Research Reports 526, Peterborough: English Nature.

Harrison, C. and Burgess, J. (1994) 'Social constructions of nature: a case study of conflicts over the development of Rainham Marshes SSSI', *Transactions of the Institute of British Geographers*, vol. 19, 291–310.

Harrison, C. and Davies, G. (2002) 'Conserving biodiversity that matters: practitioners' perspectives on brownfield development and nature conservation in London', *Journal of Environmental Management*, vol. 65, 95–108.

Harrison, C. and Haklay, M. (2002) 'The potential of public participation geographic information systems in UK environmental planning: appraisals by active publics', *Journal of Environmental Planning and Management*, vol. 45, 841–63.

Harrison, C., Burgess, J. and Clark, J. (1998) 'Discounted knowledges: farmers' and residents' understandings of nature conservation goals and policies', *Journal of Environmental Management*, vol. 54, 305–20.

Harrison, C., Burgess, J. and Clark, J. (1999) 'Capturing values for nature: ecological, economic and cultural perspectives', in J. Holder and D. McGillivray (eds) *Locality and Identity: Environmental Issues in Law and Society*, Aldershot: Ashgate, pp. 85–110.

Harrison, C., Burgess, J., Millward, A. and Dawe, G. (1995) *Accessible Natural Greenspace in Towns and Cities: A Review of Appropriate Size and Distance Criteria*, English Nature Research Reports 153, Peterborough: English Nature.

Harrison, C., Limb, M. and Burgess, J. (1986) 'Recreation 2000: views of the country from the city', *Landscape Research*, vol. 11, 19–24.

Harrison, C., Limb, M. and Burgess, J. (1987) 'Nature in the city: popular values for a living world', *Journal of Environmental Management*, vol. 25, 347–62.

Healey, P. (1997) *Collaborative Planning: Shaping Places in Fragmented Societies*, Basingstoke: Macmillan.

Her Majesty's Government (1999) *A Better Quality of Life*, London: The Stationery Office.

Her Majesty's Government (2005) *Securing the Future: Delivering UK Sustainable Development Strategy*, London: The Stationery Office.

Hillier, J. (2000) 'Going round the back? Complex networks and informal action in local planning processes', *Environment and Planning A*, vol. 32, 33–54.

Innes, J. and Booher, D. (2000) 'Indicators for sustainable communities: a strategy building on complexity theory and distributed intelligence', *Planning Theory and Practice*, vol. 1, pp. 173–86.

In't Veld, R.J. (ed.) (2000) *Willingly and Knowingly*, Utrecht: Lemma Publishers.

Jacobs, M. (1994) 'The limits to neoclassicism: towards an institutional environmental economics', in M. Redclift and T. Benton (eds) *Social Theory and the Global Environment*, London: Routledge, pp. 67–91.

Joss, S. and Durrant, J. (eds) (1995) *Public Participation in Science: The Role of Consensus Conferences in Europe*, London: Science Museum.

Levett, R. (2006) 'Quality of Life Assessment: a tool whose time has come?', *Ecos*, vol. 27, 61–7.

Lowndes, V., Pratchett, L. and Stoker, G. (2001a) 'Trends in public participation. Part 1: Local government perspectives', *Public Administration*, vol. 78, 205–22.

Lowndes, V., Pratchett, L. and Stoker, G. (2001b) 'Trends in public participation. Part 2: Citizens' perspectives', *Public Administration*, vol. 79, 445–55.

Marren, P. (2002) *Nature Conservation*, London: HarperCollins.

Munton, R.J.C. (2003) 'Deliberative democracy and environmental decision-making', in F. Berkhout, M. Leach and I. Scoones (eds) *Negotiating Environmental Change*, Cheltenham: Edward Elgar, pp. 109–36.

Murdoch, J. and Abram, S. (2002) *Rationalities of Planning*, Aldershot: Ashgate.

National Audit Office (2006) *Audit of Greenspaces*, London: National Audit Office.

Office of the Deputy Prime Minister (2002a) *Planning Policy Guidance 17: Open Space, Sport and Recreation*, London: Office of the Deputy Prime Minister.

Office of the Deputy Prime Minister (2002b) *Assessing Needs and Opportunities: A Companion Guide to PPG17*, London: Office of the Deputy Prime Minister.

Owens, S. (2005) 'Making a difference? Some perspectives on environmental research and policy', *Transactions of the Institute of British Geographers*, vol. 30, 287–92.

Phillips, A. and Ashcroft, P. (1987) 'The impact of research in countryside recreation policy', *Leisure Studies*, vol. 6, 315–28.

Ratcliffe, D. (1977) *A Nature Conservation Review: The Selection of Biological Sites of National Importance to Nature Conservation in Britain*, 2 vols, Cambridge: Cambridge University Press.

Rydin, Y. (2003) *Urban and Environmental Planning in the UK*, 2nd edn, Basingstoke: Macmillan.

Rydin, Y., Holman, N., Hands, V. and Sommer, F. (2003) 'Incorporating sustainable development concerns into an urban regeneration project: how politics can defeat procedures', *Journal of Environmental Planning and Management*, vol. 46, 545–61.

Stoker, G. (1999a) *The New Politics of Local Governance*, London: Macmillan.

Stoker, G. (1999b) *The New Management of Local Governance*, London: Macmillan.

United Kingdom Man and the Biosphere Urban Forum (2006) www.ukmaburbanforum.org.uk.

Urban Green Spaces Task Force (2002) *Green Spaces, Better Places: Final Report of the Urban Green Spaces Task Force*, London: Department of Transport, London and the Environment.

Urban Task Force (1999) *Towards an Urban Renaissance: Final Report of the Urban Task Force*, London: Spon.

9 Single farm payments, diversification and tenant farming in England

Brian Ilbery, Damian Maye, David Watts and Lewis Holloway

Introduction

Current reforms to the European Union's Common Agricultural Policy (CAP) raise important questions about the future structure and nature of agricultural production in the United Kingdom (see Policy Commission on Farming and Food 2002; Lobley and Potter 2004; Robertson 2004 and Rutherford 2004 for reviews). However, the position of tenant farmers in this changing agronomic climate is far from clear. The main reason for such uncertainty is the fact that CAP reforms are likely to impact on tenant farms in different ways according to the type of tenancy, the nature of the landlord–tenant relationship and the geographical location of the farm enterprise. Tenant farmers, of different types, often have less freedom of choice in their decision-making compared with owner-occupiers, especially in relation to their farming and diversification activities. Past research on the Farm Diversification Grant Scheme by Ilbery and Bowler (1993), for example, showed that many tenant farmers on so-called full agricultural tenancies (FATs) were restricted in their diversification activities by contractual arrangements with landlords.

However, in 1995 the Agricultural Tenancies Act created farm business tenancies (FBTs) as a new type of landlord–tenant agreement. FBTs were designed with three interrelated objectives in mind (Whitehead *et al.* 1997): first, to encourage more land to be made available for farming under tenancy agreements; second, to provide opportunities for new entrants to the farming sector; and third, to promote efficiency in agricultural land use. Overall, FBTs were designed to provide more flexibility in landlord–tenant agreements and to allow greater opportunities for farmers to engage in diversification activities and to expand and/or restructure their businesses. Nevertheless, most FBT agreements have covered relatively small areas of land, for relatively short periods of time, and Whitehead *et al.* (2002) indicated that around two-thirds of FBT landlords were private individuals, the rest tending to be 'traditional' institutions such as the Church of England, the National Trust and local authorities. While they increase flexibility in the tenancy sector, there is little evidence that FBTs have encouraged tenant farmers to engage in diversification activities. Of those farmers farming exclusively under an FBT, only 13 per cent had become involved in diversification activities (ibid.).

The Mid-Term Review of the CAP in 2003, and single farm payments (SFPs) in particular, could have a significant impact on landlord–tenant relationships, and there is already concern that tenant farmers may struggle to hold on to their subsidy entitlement under SFP if landlords attempt to bring existing farm tenancies to an end as a lever to extract concessions from tenants on any SFP entitlement. The situation is made more complex by the regionalized system of payments under SFPs and the spatial dynamics of farm tenancy in England. Likewise, modulation and the decoupling of subsidies from agricultural production could either open up opportunities for tenant farmers to diversify their activities or make it more difficult for them to respond to the realities of market demand.

However, before one can begin to assess the impacts of CAP reform and SFPs on the opportunities or otherwise for farm diversification by different types of tenant farmers, there is a need for research to examine both past and intended diversification activities by tenant farmers. This is important because the 2004 UK Agricultural Census showed that 28 per cent of all holdings and 35 per cent of agricultural land in England were farmed and held under a tenancy agreement. Geographically, the area of land farmed under tenancy agreements in England shows a bias towards the northern counties, especially Northumberland. There are also marked spatial differences in the proportions of tenanted land farmed under FATs and FBTs to the north and south of a line drawn approximately between Liverpool and Ipswich. To the north of this line, the area of land under tenancy agreements is biased towards FATs (Figure 9.1), while to the south of the line FBTs are relatively more significant. These differences probably relate to the slower rates of change in landownership in northern regions.

The overall purpose of this chapter, therefore, is to assess the potential impacts that the introduction of SFPs and other CAP reforms may have on tenant farmers in England, focusing in particular on their diversification activities. More specifically, it analyses current tenant farmer diversification activities and examines future intentions towards diversification in different agricultural regions. The main empirical data drawn upon in this chapter were derived from a Department for Environment, Food and Rural Affairs (DEFRA)-funded research project conducted by the authors.[1] Ownership and property relations are important elements of rural restructuring in the United Kingdom (Lobley and Potter 2004; Winter, this volume, Chapter 4). This was recognized a long time ago in the seminal work of Newby *et al.* (1978). However, since their original study on farm tenancy and property relations, relatively little has been published on this key topic (for exceptions, see Whatmore *et al.* 1990; Marsden *et al.* 1993; and Spencer 2000). Thus the next two sections outline the main dimensions of the 2003 CAP reforms and the nature of farm diversification, before the rest of the chapter examines the potential impacts of these reforms on the diversification activities of different types of tenant farmers across England.

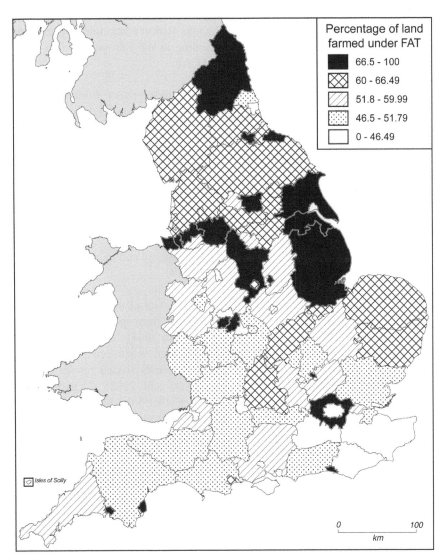

Figure 9.1 Proportion of tenanted land under full agricultural tenancies (FATs) in English counties and unitary authorities.

Note
This work is based on data provided through EDINA UKBORDERS with the support of the ESRC and JISC, and uses boundary material that is Crown copyright.

CAP reforms and single farm payments

The 2003 Mid-Term Review of the CAP could mark the beginning of profound changes in the policy environment for farming in the European Union (EU). There are five key elements of the reformed CAP:

- a single farm payment for EU farmers, independent from production (limited coupled elements may be maintained to avoid abandonment of production);
- the linking of this SFP to the requirement to keep all farmland in good agricultural and environmental condition (cross-compliance), and to respect food safety and animal welfare standards;
- a strengthened rural development policy with more money (the second 'pillar' of the CAP);
- a reduction in direct payments (modulation) for bigger farms to finance the new rural development policy;
- new measures to promote the environment, quality and animal welfare, and to help farmers meet EU production standards.

If it is examined in more detail, the most important element of the reformed CAP is the decoupling of 11 farm subsidy schemes from agricultural production and their consolidation into SFPs (Ward and Lowe 2004). English farmers will receive flat-rate 'regionalized area payments', phased in between 2005 and 2012 on a sliding scale and displacing 'historic' payments (based on the average subsidy received by each farm between 2000 and 2002) (Department of Environment, Food and Rural Affairs 2004: 1; Rutherford 2004: 25). The move from 'historic' to flat-rate payments will alter the amount of subsidy received by many farms. It is for this reason that they will be phased in and 'regionalized'. For the payment of SFPs, English farmland is classified as belonging to one of three 'regions': moorland within Severely Disadvantaged Areas (SDAs); other land in SDA; and non-SDA. English SDAs were defined in statutory maps published in 1992. Together with Disadvantaged Areas (DA), they form the Less Favoured Areas (LFA). The phasing in and 'regionalization' of SFP are intended to mitigate, over time and space, the effects of the transfer of subsidy away from some types of farm and towards others. The Department for Environment, Food and Rural Affairs (2004: 2) envisages that some 13 per cent of total subsidy will be redistributed under the flat-rate SFPs: hence the phasing of the withdrawal, until 2012, of 'historic' payments. The 'regionalization' of England will 'ring-fence' subsidy payments within moorland SDAs, other SDAs and non-SDAs, so that a 'large redistribution from the lowland to the hills is avoided' (ibid.: 2).

Making SFP conditional on cross-compliance seems calculated to mitigate the deleterious environmental consequences of agricultural intensification. As such, it would appear to affirm the EU's commitment to a 'multifunctional' model of agriculture, which posits that farming, in addition to producing food, also sustains rural landscapes, protects biodiversity, generates employment and contributes to the viability of rural areas (Potter and Burney 2002; Ward and Lowe 2004; Potter and Tilzey 2005). Indeed, the rewarding of farmers for positive environmental management is made possible by the increased modulation from Pillar 1 of the CAP (the SFP) to Pillar II (Rural Development Regulation). This extra money will fund two new agri-environmental schemes: Entry Level Stewardship (ELS) and Higher Level Stewardship (HLS). ELS is open to all farmers and is intended to deliver real environmental benefits over and above the levels that will be required under cross-compliance measures.

At this stage, it is worth reiterating the wider context of the 2003 CAP reforms: 'the primary goal, and greatest success, from the EU's and UK government's perspective, was securing progress on the liberalization of global trade' (Rutherford 2004: 29). The reforms, therefore, are in line with the government's neo-liberal approach to the food sector in general (Barling and Lang 2003). Such an approach is likely to have two main consequences for England's farming sector. First, 'emphasis on the need for farmers to be able to compete in a liberalized global market seems to place greater emphasis worldwide on the continuation of productivist principles' (Evans *et al.* 2002: 316). Thus, the productivist trend of fewer, larger, more intensive and specialized farms looks set to continue. Second, this neo-liberal approach is unlikely to reduce the social and economic inequalities that already exist between marginal and/or remote rural areas and those that have been, or can be, incorporated into productivist agriculture. If anything, they may increase. However, it is too early to judge the extent to which either of these developments will accelerate in England.

Farm diversification

A recent benchmarking study on farm diversification by Exeter University noted that there remain 'considerable definitional difficulties surrounding the term' (Centre for Rural Research 2002: 12). However, the report goes on to claim that '[m]ost commentators would now agree that diversification encompasses business activities that are run on the farm or are dependent on farm-based land and capital assets' (ibid.: 12). The Exeter study considered farm diversification as a subset of farm household pluriactivity (ibid.: 12). Pluriactivity is usually defined as 'farming in conjunction with another gainful activity, whether on- or off-farm' (MacKinnon *et al.* 1991: 59, cited by Evans and Ilbery 1993: 949; see also Bowler *et al.* 1996: 286–7). The distinction between pluriactivity and farm diversification would therefore appear to be clear-cut: the former encompasses all gainful activity undertaken in addition to 'conventional' farming; the latter includes only such activity carried out on the farm. However, matters are complicated by the Exeter study's inclusion, under diversification, of 'business activities ... dependent on farm-based land and capital assets' (Centre for Rural Research 2002: 12). Thus, some off-farm gainful activities, notably agricultural contracting services and hirework, are defined as diversification. There is a good reason for considering such activities as farm diversification. Both involve the gainful, off-farm use of assets that are used for 'conventional' agricultural production on the farm that owns or hires them. This enables the concept of diversification to encompass all the economies of scope[2] that the farm business exploits, including the agricultural knowledge and capacities of its personnel and equipment.

For this research, a tentative division was made between non-conventional farming activities and non-farming activities. While the former include specialist products (livestock, crop and organic), food processing and direct marketing, the latter include sports/leisure facilities, accommodation services and hire/contract services. These are listed in Table 9.1. It is recognized that there are problems with such a restricted definition. For example, excluding off-farm, non-agricultural gainful activities from studies of farm diversification may under-report the contribution of female members to the farm household (Evans and Ilbery 1996). Likewise, it may also create a farm size bias, because larger farms are more likely to diversify than smaller ones (Ilbery 1991; Ilbery and Bowler 1993). This is because '[f]amilies on small farms whose income needs may well be greater, have surplus labour rather than land or capital to deploy' (Gasson 1988: 179). More recently, McNally's (2001: 251) longitudinal analysis of the Farm Business Survey (FBS) revealed that 'average net farm incomes are higher for diversifiers than for the [FBS] sample as a whole'. She also found that the leasing of land or buildings is one of the most popular forms of diversification, a finding confirmed by Lobley and Potter (2004).

Several authors have argued that diversification rates tend to be lower for tenanted than for owner-occupied farms because of restrictions placed on non-agricultural activities by tenancy agreements (Bateman and Ray 1994: 8; Ilbery 1991: 216; McNally 2001: 255). However, the recent Exeter University benchmarking study contradicts this suggestion. It found that wholly owner-occupied holdings are less likely than average to be involved in farm diversification (at 54.5 per cent), with the opposite being the case for wholly tenanted holdings (at 66.4 per cent) (Centre for Rural Research 2002: xv). Thus, while diversification on tenant farms is undoubtedly widespread (ibid.: 34), disagreement remains over the extent to which tenant farmers are disadvantaged by having to seek landlord consent in order to diversify.

Table 9.1 Farm diversification activities

Non-conventional farming	*Non-farming*
Specialist products: • Livestock products • Organic products • Crop products • Contracting Food processing and direct marking: • Food preparation/processing • Direct marketing	• Sports/leisure facilities • Accommodation services • Hire/contract services • Hire/leasing • Others

A survey of tenant farmers

In order to examine the three interrelated themes of farm tenancy, farm diversification and the potential impacts of the 2003 CAP reforms, empirical data were collected in two main stages. The first phase comprised a national postal survey of different farm tenant types throughout England. In order to provide a representative sample of the 44,206 eligible tenanted holdings across England, the sample was selected according to three criteria:

• Government Office Region (GOR);
• Type of tenure, in terms of the proportion of rented farmland. The three categories used were wholly tenanted (100 per cent), mainly tenanted (75–99 per cent) and partly tenanted (10–74 per cent);
• Designated area status, in terms of whether or not holdings with tenanted land are located in a Less Favoured Area (LFA); LFAs were further divided into Disadvantaged (DAs) and Severely Disadvantaged Areas (SDAs).

Finally, the sample of 3,360 tenant farms was selected according to the proportions of eligible holdings found against each of the three sampling criteria.

The postal questionnaire was designed to analyse both the *current* nature of farm tenancy and farm diversification activities, and how the new changes to the CAP might impact on tenancy and especially *future* diversification plans. Emphasis within the questionnaire moved from obtaining basic profile information about the farm (e.g. in terms of farm tenure, business structure and type of tenancy) to considering diversification activities, the links these have with tenancy agreements, and how such arrangements might change in response to the current policy reforms. The overall response rate, at 11.1 per cent, was low for three main reasons: first, the timing of the survey at a busy time in the farming calendar (August); second, the detailed nature of the questionnaire; and third, the real uncertainty surrounding the impacts of SFPs on the future of farming in the United Kingdom. Response rates tended to be higher in northern and eastern regions than in southern and western regions, and from wholly and mainly tenanted rather than partly tenanted holdings; likewise, a higher response was obtained from SDA than non-SDA areas, possibly reflecting greater concerns about CAP reforms among tenant farmers in SDAs. Despite the relatively low response rate, the quality of information provided in the fully completed questionnaires was very high.

Over 40 per cent of those completing the questionnaire indicated a willingness to take part in a follow-up telephone interview. Thus, the second phase of data collection included telephone interviews with 52 tenant farmers, drawn according to tenure type (i.e. wholly, mainly and partly tenanted) and SFP 'region', and 17 landlords, selected with help from the Country Land and Business Association (CLA). The latter did not necessarily have direct involvement with the 52 tenant farmers. This phase of the research project was interested in four main topics: landlord–tenant relations; diversification; impacts of CAP reform; and the future for the tenanted sector. The following analysis draws primarily upon the postal questionnaire survey, but also makes use of the tele-phone interviews where appropriate.

Farm tenure and current diversification activities

This section outlines the main characteristics of the different types of tenant farmer and landlord among the sample population, before examining the current nature of farm diversification activities. A key finding from the postal question-naire survey sample was the lack of statistically significant relationships between tenancy and either farm size or farm type. However, the sample showed a slight over-representation of partly tenanted farms and a small under-representation of wholly tenanted farms. Interestingly, FATs remain the most numerous form of tenancy, accounting for nearly two-thirds of agreements where farmers rent from a single landlord; this contrasts with a figure of just over 25 per cent for FBTs. Nevertheless, this is likely to change because, within five years, FBTs will prob-ably be the most numerous type of agricultural tenancy in England. The survey sample was also biased towards larger farms and mixed, dairy and arable enter-prises. Over half of all respondents have farms of over 100 hectares, compared to 14 per cent of farmers nationally. Interestingly, wholly tenanted farms account for a majority (54 per cent) of responding farms of less than 50 hectares. It is perhaps not surprising that mixed, dairy and arable farms are over-represented, because it is these farming types that DEFRA predicts will lose the most under the flat-rate – Regionalized Area Payment – SFP.

Approximately 40 per cent of tenanted farms rent land from more than one landlord, with a small percentage (2.3 per cent) claiming to rent from four or more. Table 9.2 indicates that private owners are easily the most common type

Table 9.2 Main landowner types among tenant farmers

Landowner type		Landowner 1	Landowner 2	Landowner 3	Total
Private owner	Number	89	58	33	180
	Percentage	35.7	61.1	71.7	46.2
Landed estate	Number	63	9	0	72
	Percentage	25.3	9.5		18.5
Institution (incl. Crown)	Number	36	10	4	50
	Percentage	14.5	10.5	8.7	12.8
Other	Number	25	9	5	39
	Percentage	10.0	9.5	10.9	10.0
Local authority	Number	19	2	2	23
	Percentage	7.6	2.1	4.3	5.9
Family member	Number	15	7	2	24
	Percentage	6.0	7.4	4.3	6.2
Financial institution	Number	2	0	0	2
	Percentage	0.8			0.5
Totals		249	95	46	390

of landlord, followed by landed estates and institutions. Significantly, private owners are also the most numerous type of landlord from whom farmers rent land in addition to their main holding. However, they account for only 27 per cent of wholly tenanted holdings rented from a single landlord and are easily surpassed by landed estates in this category. The fact that landed estates account for the largest number of wholly tenanted farms with a single owner (36 per cent) reflects the continued resilience of this traditional (mainly FAT) segment of the let sector (see also Spencer 2000 in relation to landholding patterns among the Oxford colleges).

According to their own, lay definitions, fewer than half the postal respondents (43 per cent) stated that they engage in farm diversification activities. This contrasts with the 70.3 per cent who undertake one or more of the activities listed in Table 9.1. The main reason for this discrepancy lies in the fact that many farmers do not consider farm contracting to be a form of diversification, even though it is the single most popular form of diversification among the survey sample (Table 9.3). There are two main reasons behind this. First, contracting is seen as a traditional activity that complements their farming activities. For many, small-scale contracting is simply a case of farmers helping out their neighbours and something that has gone on for years. Second, contract farming

Table 9.3 Types of diversification activity undertaken

(a) Main categories of diversification

Type of diversification activity	Number of respondents undertaking it	Percentage of all respondents (n = 266)	Percentage of diversified respondents (n = 187)
Hire/contract services	94	35.3	50.3
Sports/leisure facilities	75	28.2	40.1
Accommodation services	56	21.1	29.9
Specialist products	53	19.9	28.3
Food processing/direct marketing	52	19.5	27.8
Other products/services	12	4.5	6.4

(b) Most popular individual diversification activities

Diversification activity	Percentage of respondents undertaking it
Agricultural contracting	24.1
Livery	14.3
Shooting	10.2
Year-round domestic accommodation	9.4
Commercial building rental/leasing	7.5
Bed and breakfast accommodation	6.8
Vehicle storage/parking	5.6
Machinery hire	5.6
Short-term self-catering accommodation	5.6

is often an expansion of their core business and not a diversification activity. Yet, in contrast, some respondents regarded membership of an agri-environmental scheme as a diversification activity. Such vagueness over the concept of farm diversification also pervaded the telephone interviews with selected farmers and landlords.

Despite difficulties of definition, the survey reported high rates of farm diversification among farms with tenanted land; at 70.3 per cent, this exceeds the 65.7 per cent found for wholly and mainly tenanted farms by the Exeter University benchmarking study (Centre for Rural Research 2002). Not only have seven out of ten tenant farmers diversified, but over 60 per cent of these undertake two or more diversification activities; indeed, a not insignificant 9 per cent of diversifiers have six or more separate activities. At 2.5, the mean number of diversified activities per farm would lend some support to Carter's (1999: 427) notion that a proportion of 'farms are emerging as centres of mixed entrepreneurial talent' and making an important contribution to rural business development (Jarvis *et al.* 2002). In terms of the broad categories of farm diversification, hire and/or contract services emerge as the most popular and are undertaken by half of those farmers who have diversified. This is followed by sports and leisure facilities (40 per cent) and three other categories that range in popularity from 28 to 30 per cent: accommodation services, specialist products and food processing/direct marketing (Table 9.3). These results are in line with previous findings (McInerney *et al.* 1989; McNally 2001; Lobley and Potter 2004).

As was suggested earlier, agricultural contracting is much the most important form of farm diversification, being undertaken by one in four respondents; this is followed by livery (14 per cent) and shooting (10 per cent). Many of these activities are consistent with farmers aiming to maximize their economies of scope by exploiting existing assets (land, buildings and equipment) and knowledge. Indeed, five of the top nine activities involve relatively low levels of labour input and relate to what Ilbery (1991) termed 'passive' diversification. Also, much of the livery provision is on a do-it-yourself basis in order to minimize the farm household's labour input. Likewise, in agreement with earlier studies by Gasson and Winter (1992) and by Evans and Ilbery (1996), farmers' spouses are often responsible for the provision of bed and breakfast, and self-catering accommodation.

The type of farming practised seems to bear little resemblance to the different diversification activities, although understandably there is a tendency for the relatively small number of horticultural holdings to concentrate on adding value to their produce through processing and/or direct marketing. However, farm size does have a major impact on farm diversification, with farms over 50 hectares accounting for 80 per cent of all diversified farms in the survey. These larger farms account for 85 per cent and 87 per cent of farms that have diversified into the provision of hire/contract and sports/leisure services. Interestingly, farms located in SDAs are less likely to diversify than those located outside (66 per cent against 71 per cent). However, this figure rises to nearly 80 per cent for farms with land in moorland SDAs, where farm-based tourism is often predominant (Ilbery *et al.* 1998).

A key focus of the research was the nature of landlord–tenant relationships, especially in terms of how they impact on diversification activities. While just over half the farmers with tenanted land thought that their landlord would support some form of diversification, nearly 60 per cent reported that their landlord had expressed no view on this topic. Interestingly, the proportion of landlords thought to favour diversification decreases as the amount of tenanted land decreases. In more detail, many farmers perceive that their landlord would not approve any activity that would be likely to have a negative impact on the character of the estate or its physical environment. Others felt that permission to diversify (except possibly for agricultural contracting) would be granted on the basis of some increase in rent. Indeed, telephone interviews with landlords confirmed that rent is often a cause of considerable friction, especially because many farmers will not acknowledge that without the landlord's resources (usually a building), diversification could not take place. As a possible consequence, the research indicated that there has been only a limited amount of investment in farm diversification by landlords. It is not surprising, therefore, that, where possible, farmers with tenanted land prefer to diversify on land that they own, with many using their rented land solely for farming. In this way, they benefit from any potential capital appreciation. This is an important point because, as the amount of rented land increases, the amount of diversification may decrease, especially as much 'new' rented land is small parcels let on short FBTs.

Overall, a number of farmers with tenanted land claimed that their tenancy agreements restrict their ability to diversify into non-farming activities. In an attempt to ease this difficulty, the Tenancy Reform Industry Group (TRIG) produced a code of good practice (published by DEFRA) to assist landlords and tenants in negotiating either party's proposals for gainful activities outside the traditional agricultural enterprises stipulated by most tenancy agreements (Tenancy Reform Industry Group n.d.). In 2005 this code was supplemented by a voluntary, non-binding arbitration scheme operated by the Royal Institute of Chartered Surveyors for cases where tenants and landlords cannot resolve their differences. However, while the telephone interviews with landlords showed some awareness of the TRIG code of practice, 37 of the 52 tenant farmers interviewed (71 per cent) claimed not to have heard of it.

This section has raised a number of key points about tenancy and farm diversification. First, there is no clear relationship between farm tenure and either farm size or farm type. Second, private owners are by far the most numerous type of landlord in England. Third, there is considerable divergence between what academics and tenant farmers consider to be diversification. Fourth, while high rates of diversification exist among tenant farmers in England (70 per cent), there is a clear bias towards diversification on larger farms. Lastly, farmers with tenanted land prefer to diversify on land that they own.

CAP reforms: impacts and future scenarios

The chapter now turns to consider the potential impacts of SFPs on tenanted farm businesses in England. This section examines tenant farmers' views on whether or not landlord attitudes towards diversification are likely to have changed in the light of SFPs and any future plans for diversification they have, before looking more generally at possible farm business trajectories over the next two or three years. Given the scheduling of the survey work in 2005, when farmers did not know either the level or the timing of the SFPs, it is not surprising that responses to the CAP reforms were dominated by uncertainty (49 per cent) and negativity (36 per cent). Fewer than 6 per cent of the postal survey respondents thought that the CAP reforms would have a positive impact on their business. This concern characterizes all agricultural regions, but varies according to tenure type. Thus, while a relatively large proportion of partly tenanted farmers (57 per cent) remain uncertain about CAP reforms and SFP, a significant number of mainly tenanted farmers (32 per cent) are quite negative; this is particularly the case among tenants of landed estates, where an older age structure may lead to greater resistance to change. In contrast, farmers who rent land from private landlords are less concerned, and many feel that the 2003 CAP reforms will not affect them in any substantial way.

Irrespective of these general attitudes, the vast majority of tenants (86 per cent) believe that they will receive the SFP; nevertheless, a significant proportion (around 30 per cent) are not sure what will happen to the SFP entitlement when their tenancy agreement expires. About half of the farmers have discussed the CAP reforms with their landlord(s). Understandably, the level and timing of SFP payments dominated discussions. Diversification was discussed in less than 5 per cent of cases. This does not mean that farm diversification is being neglected in discussions between tenants and landlords; it is just that more urgent matters surrounding SFPs are likely to have pushed future investment in diversification into the background. Significantly, a large proportion of tenant farmers (36 per cent) do not know whether the CAP reforms have changed their landlords' attitudes towards diversification; this reinforces the sense of uncertainty that pervades the tenanted farming sector. A further 31 per cent thought that landlord attitudes would not be changed by the CAP reforms, whereas 24 per cent felt that landlords would now be more willing to permit diversification.

When tenant farmers were asked directly about whether the introduction of the SFP would influence their future plans for diversification, a clear distinction emerged between those who had and those who had not already diversified. For more than a third of the former, the introduction of the SFP is likely to influence their future diversification plans (Table 9.4). In contrast, and significantly, nearly 80 per cent of those who have not already diversified stated that introduction of the SFP will have little impact on their future plans to do so. However, both the postal and the interview stages of the research highlighted other factors, apart from SFPs, that might restrict future diversification plans. These included a lack of time and finance, poor location (away from passing trade), a perceived satu-

Table 9.4 Single farm payments (SFPs) and future diversification plans

Has SFP influenced current or future diversification plans?		Has the farm already diversified?	
		No	Yes
No	Number	41	77
	Percentage	78.8	56.6
Yes	Number	8	47
	Percentage	15.4	34.6
Not yet known	Number	3	12
	Percentage	5.8	8.8
Totals		52	136

rated market and a type of farming that does not lend itself to adding value through processing and/or direct marketing. For many tenant farmers, especially in upland areas, the generation of income from off-farm activities rather than on-farm diversification is a better future possibility.

The sheer scale of concern over the SFP and tenancy arrangements undoubt-edly affected possible future farm business trajectories (Table 9.5). Dominant among the different pathways is the decision to carry on farming as before and wait until things settle down before deciding on longer-term plans (42 per cent). One in five intend to pursue economies of scale through increasing the size of business and/or expanding production, and another 16 per cent aim to exploit economies of scope through either diversifying for the first time or expanding

Table 9.5 Future farm business trajectories

Plans for the future of the farm	Number of ticks	Percentage	Cumulative percentage
Carry on farming as before	143	37.8	
Will let things settle down before deciding	14	3.7	41.5
Increase size of farm business	53	14.0	
Intensify production	26	6.9	20.9
Diversify farm business	30	7.9	
Further diversify farm business	30	7.9	15.8
Downsize farm business	43	11.4	
May quit farming	8	2.1	13.5
Go into partnership with other farm(s)	19	5.0	
Other/don't know	12	3.2	
Total	378		

current activities. Of course, these options are not mutually exclusive, and a number of farmers are thinking of following more than one pathway. A not insignificant proportion (14 per cent) feel that they will either downsize their business or quit altogether. This may reflect the general pessimism surrounding SFPs, a desire to retire or a wish not to engage in a more market-oriented agriculture or agri-environmental schemes.

However, if one in seven tenant farmers decide to downsize or give up farming, it is unlikely that the land they currently farm will go out of production. Results show that many farmers are looking to expand their operations, and the Tenant Farmers Association claims that there are plenty of aspiring new tenants wanting to get into farming. Although the possible future pathways do not vary according to farm type, there are some interesting differences in terms of tenure. For example, the option of 'carrying on as before' is dominated by wholly tenanted farmers, whereas the intention to diversify or further diversify is higher for partly tenanted farmers. Farmers in this group are also most likely to reduce the size of their business in the future. It seems that wholly tenanted farmers regard themselves as less likely to change or diversify in the future than tenant farmers as a general group. Of course, this could reflect the Centre for Rural Research (2002) finding that this group is already the most diversified.

It is interesting to 'match' these future possibilities with an important view to emerge from the telephone interviews with tenant farmers and landlords and their agents: that, in line with developments in farming generally, there will be a decline in the number of wholly tenanted farms, but not a decline in the amount of land available. This is at odds with the findings of a recent report by the Central Association of Agricultural Valuers (2006), which suggested that the amount of let land in England fell by 2.5 per cent in 2004 and 4.5 per cent in 2005 – after eight years of successive growth following the introduction of FBTs in 1995. The report attributes this reversal to landowners' desire to regain occupation of their land in order to obtain entitlement and thus SFPs. Several interviewees were aware of estates taking the opportunity, when a lease expires or is surrendered, of letting the land in one or more parcels to existing tenants and/or other farmers, and either leasing or selling the farmhouse for residential use. Significant increases in house prices, combined with poor returns from agriculture, mean that in many areas farmhouses are now worth more as private residences than they are as part of an equipped holding. Agricultural rents now form only a relatively small proportion of the overall income for many estates, including those of 'traditional' landowners such as the Church Commissioners and the colleges of Oxford University (Spencer 2000: 301). As one interviewee suggested, 'agricultural land is now no more than an appendage to where the real value lies on this estate: the property'.

It follows from this that two particular sub-groups of the let sector – traditional tenants and new entrants – are likely to fare the worst over the coming years. In the view of many interviewees, land that becomes available for rent, whether or not it is new to the let sector, is likely to go to existing farmers who are looking for ways of spreading their fixed costs over a larger area. Owner-occupiers are probably in a stronger position to bid for such land, for three main reasons. The first is the additional property rights that landownership confers (Whatmore *et al.* 1990: 240). For instance, having ownership rights over land provides a source of collateral when borrowing to finance future investment. Second, tenants often have less flexibility, owing to conditions in their tenancy agreements, over the ways in which they can farm. This means that they may be less able to diversify on their 'core' holding, thereby depriving them of opportunities to generate additional income with which to finance new investment. Third, tenant farmers have to pay council tax on property that they do not own, with the result that such payments are not offset by increases in the value of the houses they inhabit. Thus, tenant farmers may lose out to owner-occupiers seeking to expand their businesses. These difficulties are likely to be magnified for aspiring tenants to the let sector, who will, in addition, lack the track record in farming that owners and lenders are increasingly likely to require as evidence that applicants can generate a commercial rate of return from farming. Indeed, the lack of new tenants to the tenanted sector was acknowledged to be a problem by several interviewees. If these predictions are fulfilled, it is possible that the tenanted sector, as it is currently understood, may all but disappear. In cases where entitlement to claim the SFP is being held by the landlord, rents (especially for bare land) may in the future fall towards zero, with economic value released by altering tenancy arrangements to 'free up' other assets, especially farm and residential buildings.

As a summary to this section, a number of important observations can be made. First, there is real uncertainty about the potential impacts of the 2003 CAP reforms and SFPs, especially among mainly tenanted farmers and on traditional landed estates. Second, the vast majority of tenant farmers believe that they are entitled to SFPs. Third, in terms of SFP impact, there is a clear distinction between those tenant farmers who have and have not already diversified, with the latter still unlikely to diversify in the future in response to the SFP. Fourth, the most likely future farm business trajectory is to 'carry on farming as before', with far fewer tenant farmers seeking to pursue either economies of scale through farm business expansion or economies of scope through diversification activities. Finally, the importance of property and landownership indicates that owner-occupiers are in a much stronger position than aspiring new tenant farmers to bid for any land that becomes available for rent in the future.

Conclusions

The 2003 CAP reforms and introduction of SFP are almost certain to have a major impact on agricultural restructuring in England over the next ten to 20 years. As part of the government's increasingly neo-liberal approach to the food sector in general, the move towards more 'market' orientation in agriculture could lead to increasing spatial disparities between core and marginal farming regions, with most food being produced under productivist principles in the former and little being produced in the latter. The position of tenant farmers within this restructuring process is far from clear, and survey results indicate that the sector is characterized by high levels of uncertainty and negativity in terms of how the CAP reforms, and SFPs in particular, will impact on their farm business behaviour. Tenant farmers (and landlords) are currently more concerned about levels of SFP and issues surrounding entitlement than they are with any future investment decisions, including those related to diversification activities. Significantly, the majority of tenant farmers who have not already diversified said they are unlikely to consider diversification activities in the future. Nevertheless, the sheer scale of uncertainty and negativity surrounding the CAP reforms and SFPs means that it is too early to fully evaluate the opportunities for, and constraints upon, future diversification possibilities among different types of tenant farmers in different regions in England. However, there is an urgent need for ongoing research on these themes in specific geographical settings.

Although the survey set out to examine the diversification activities of tenant farmers, a key issue to emerge from it is that of who is actually diversifying. Both landlords and tenants have been under considerable financial pressure for much of the past decade, and both appear to be seeking new ways of generating new income. Thus, estate assets could, and in some cases have probably already, become objects of contestation as both owner and tenant try to secure additional income from them. Several landlords intend, when farm tenancies become vacant or up for renewal, to offer tenants bare land and to let the buildings and residential properties separately. Consolidation in the tenanted sector, therefore, is being driven not only by the economics of farming but also by the significant increase in the letting value of residential and other farm buildings. Many landlords are seeking to exploit all estate assets in order to diversify their income, and in a number of cases this is being done by removing assets such as houses and agricultural buildings from farms and letting them separately. However, the situation of renting out bare land is most likely to occur in relation to new tenants and, especially, to existing farmers taking on more land in order to spread their costs. For such tenants, the issue of diversification on their newly rented land would not arise, because they would have taken it on for purely agricultural purposes.

Linked to this, it is likely that the functional distinction between the tenanted and the owner-occupied sectors will become increasingly blurred in future years. While the legal distinction will remain, it would seem that, as farming businesses get bigger, they will tend to farm more and more land under a mixture of agreements: owner-occupation, tenancy and contract. These larger agribusinesses are likely to be hybridized as far as tenure is concerned, and, in an increasingly competitive market, tenant farmers may lose out to owner-occupiers seeking to expand their businesses. While such hybridization has been taking place for a number of years, the SFP/entitlement issue has helped to highlight this process. In turn, it could become exceptionally difficult for another key element of the let sector – new entrants – to rent land, a problem acknowledged by many interviewees. Thus, it is possible that tenure may become a means to an end rather than a defining socio-economic characteristic. This is especially the case with the possible demise of wholly tenanted farms as a distinct part of the let sector. Likewise, as farm size and capitalization continue to rise, the relative advantage accruing to owner-occupiers is likely to decline, as they will be renting more land and doing more contracting.[3]

Notes

1 Research into the potential impacts of CAP reform on the diversification activities of tenant farmers in England – baseline survey (DEFRA EPES 0405/05)
2 The definition of economies of scope used here is: 'the opportunities for those operating land-based businesses to create real economies by increasing the scope (range) and value of production and services' (Marsden *et al.* 2002: 811).
3 The authors are grateful to DEFRA for funding the project 'Research into the potential impacts of CAP reform on the diversification activities of tenant farmers in England – baseline study' (EPES 0405/05).

References

Barling, D. and Lang, T. (2003) 'A reluctant food policy? The first five years of food policy under Labour', *The Political Quarterly*, vol. 74, 8–18.
Bateman, D. and Ray, C. (1994) 'Farm pluriactivity and rural policy: some evidence from Wales', *Journal of Rural Studies*, vol. 10, 1–13.
Bowler, I., Clark, G., Crockett, A., Ilbery, B. and Shaw, A. (1996) 'The development of alternative farm enterprises: a study of family labour farms in the northern Pennines of England', *Journal of Rural Studies*, vol. 12, 285–95.
Carter, S. (1999) 'Multiple business ownership in the farm sector: assessing the enterprise and employment contributions of farmers in Cambridgeshire', *Journal of Rural Studies*, vol. 15, 417–29.
Central Association of Agricultural Valuers (2006) *Tenanted Farm Survey 2005*, www.caav.org.uk (accessed 12 July 2006).
Centre for Rural Research (2002) *Farm Diversification Activities: Benchmarking Study 2002. Final report to DEFRA*, Exeter: Centre for Rural Research, University of Exeter.
Department for Environment, Food and Rural Affairs (2004) *CAP Single Payment Scheme: basis for allocation of* entitlement, www.defra.gov.uk/farm/capreform/single-pay/overview/qa-scheme.htm (downloaded 25 July 2005).
Evans, N. and Ilbery, B. (1993) 'The pluriactivity, part-time farming, and farm diversification debate', *Environment and Planning A*, vol. 25, 945–59.
Evans, N. and Ilbery, B. (1996) 'Exploring the influence of farm-based pluriactivity on gender relations in capitalist agriculture', *Sociologia Ruralis*, vol. 36, 74–92.
Evans, N., Morris, C. and Winter, M. (2002) 'Conceptualizing agriculture: a critique of post-productivism as the new orthodoxy', *Progress in Human Geography*, vol. 26, 313–32.
Gasson, R. (1988) 'Farm diversification and rural development', *Journal of Agricultural Economics*, vol. 39, 175–81.
Gasson, R. and Winter, M. (1992) 'Gender relations and farm household pluriactivity', *Journal of Rural Studies*, vol. 8, 387–97.

Ilbery, B. (1991) 'Farm diversification as an adjustment strategy on the urban fringe of the West Midlands', *Journal of Rural Studies*, vol. 7, 207–18.

Ilbery, B. and Bowler, I. (1993) 'The Farm Diversification Grant Scheme: adoption and non-adoption in England and Wales', *Environment and Planning C: Government and Policy*, vol. 11, 161–70.

Ilbery, B., Bowler, I., Clark, G., Crockett, A. and Shaw, A. (1998) 'Farm-based tourism as an alternative farm enterprise: a case study from the North Pennines, England', *Regional Studies*, vol. 32, 355–64.

Jarvis, D., Dunham, P. and Ilbery, B. (2002) 'Rural industrialization, "quality" and service: some findings from south Warwickshire and north Devon', *Area*, vol. 34, 59–69.

Lobley, M. and Potter, C. (2004) 'Agricultural change and restructuring: recent evidence from a survey of agricultural households in England', *Journal of Rural Studies*, vol. 20, 499–510.

McInerney, J., Turner, M. and Hollingham, M. (1989) *Diversification in the Use of Farm Resources*, Report 232, Department of Agricultural Economics, University of Exeter.

MacKinnon, N., Bryden, J., Bell, C., Fuller, A. and Spearman, M. (1991) 'Pluriactivity, structural change and farm household vulnerability in Western Europe', *Sociologia Ruralis*, vol. 31, 58–71.

McNally, S. (2001) 'Farm diversification in England and Wales: what can we learn from the Farm Business Survey?' *Journal of Rural Studies*, vol. 17, 247–57.

Marsden, T.K., Banks, J. and Bristow, G. (2002) 'The social management of rural nature: understanding agrarian-based rural development', *Environment and Planning A*, vol. 34, 809–25.

Marsden, T.K., Murdoch, J., Lowe, P., Munton, R.J.C. and Flynn, A. (1993) *Constructing the Countryside*, London: UCL Press.

Newby, H., Bell, C., Rose, D. and Saunders, P. (1978) *Property, Paternalism and Power*, London: Hutchinson.

Policy Commission on Farming and Food (2002) *Farming and Food: A Sustainable Future*, London: The Stationery Office.

Potter, C. and Burney, J. (2002) 'Agricultural multifunctionality in the WTO: legitimate non-trade concern or disguised protectionism?', *Journal of Rural Studies*, vol. 18, 35–47.

Potter, C. and Tilzey, M. (2005) 'Agricultural policy discourses in the European post-Fordist transition: neoliberalism, neomercantilism and multifunctionality', *Progress in Human Geography*, vol. 29, 581–600.

Robertson, J. (2004) 'CAP reform: turning the corner?', *Ecos*, vol. 25, 48–54.

Rutherford, A. (2004) 'A hall of mirrors: reflections on the 2003 CAP reform', *Ecos*, vol. 25, 22–30.

Spencer, D. (2000) 'Pulling out of landed property: the Oxford colleges and the Church Commissioners', *Area*, vol. 32, 297–306.

Tenancy Reform Industry Group (n.d.) *Notes to Aid Tenants and Landlords on CAP Reform*, www.defra.gov.uk/farm/capreform/pubs/index.htm#tenlord (downloaded 25 July 2005).

Ward, N. and Lowe, P. (2004) 'Europeanizing rural development? Implementing the CAP's second pillar in England', *International Planning Studies*, vol. 9, 121–37.

Whatmore, S., Munton, R.J.C. and Marsden, T.K. (1990) 'The rural restructuring process: emerging divisions of agricultural property-rights', *Regional Studies*, vol. 24, 235–45.

Whitehead, I., Errington, A. and Millard, N. (1997) *An Economic Evaluation of the*

Agricultural Tenancies Act 1995: A Baseline Study, Plymouth: Department of Land Use and Rural Management, University of Plymouth.

Whitehead, I., Errington, A., Millard, N. and Felton, T. (2002) *An Economic Evaluation of the Agricultural Tenancies Act 1995*, Plymouth: Department of Land Use and Rural Management, University of Plymouth.

10 Diversification, networks and rural futures in England

Julian Clark and Alun Jones

Changing policy contexts: agricultural diversification and the European Union's Common Agricultural Policy

Agricultural diversification is pivotal to European Union (EU) efforts to recast agricultural goals for the twenty-first century. This chapter examines on-farm diversification of agricultural businesses in the context of recent change in the European Union's Common Agricultural Policy (CAP). Specifically, we consider the evidence for the role of networks in the initiation and fulfilment of diversification objectives. We evaluate the impacts of this process on agricultural businesses, territories and national public policies for the rural by drawing upon extensive empirical materials collected in England by the authors as part of the European Union-funded project Innovation, Diversification and European Agricultural Situations (IDEAS) of 1999–2003.

Exhortations to agricultural businesses to diversify are now universal: from the prime minister down, England's farmers are being encouraged to introduce alternative enterprises to enhance farm profitability and survival. But while the prevalence of the message is new, diversification as a policy goal is not. Successive UK governments have targeted agriculture since the late 1970s with a battery of incentives, each framed by different economic rationales. Given, therefore, that agricultural diversification is not a new phenomenon, and has been the focus of much research already, why examine the topic here?

We argue that there are two very good reasons for study. The first is one of policy context. Since the mid-1990s, diversification has been transformed from being an activity to augment 'farming' income (i.e. income from production of agricultural commodities) to being the mechanism for delivering a 'new' agriculture that prioritizes the many non-market functions of the sector. Second, there is the issue of approach. To date, the point of departure for many studies has been to consider diversification as a phenomenon. Our intention here is to examine it as a process. Not only does this provide insight into its 'hows' and 'whys' (in particular, how diversification is implemented by agricultural business managers) but also provides the means to reflect on changing state involvement with these businesses, the ways in which policy-makers have sought to recast agriculture in rural areas, and the contribution made by agriculture as one element in a diversified and increasingly differentiated rural economy. Each mirrors themes in Richard Munton's own research.

Many definitions of agricultural diversification exist, but researchers use it broadly to refer to 'change in the traditional services, activities and products engaged in or produced on farm holdings, usually in response to changing policy, political, market and farm household influences' (Clark and Jones 2003: 5).[1] Arguably the most important of these influences in the United Kingdom is the shifting policy context of the CAP. While the CAP still embodies productivist values by offering financial support for agricultural commodities under its 20 or so market regimes (Short, this volume, Chapter 2), a new impetus became evident in 1999 with the introduction of the Rural Development Regulation (RDR). European policy-makers have sought nothing less than to chart a new course for agriculture with this measure (indeed, the then European Commissioner for Agriculture and Rural Development, Franz Fischler, underlined the RDR's importance by describing it as the CAP's 'second pillar', a phrase that has stuck). Importantly, the RDR provides funding nationally and at state level for farmers to diversify their businesses. In doing so, it is beginning to change the long-established choreography of 'economic signals' to agricultural businesses – signals that have encouraged food production for much of the past 50 years.

The late 1990s also saw 'multifunctionality' added to the CAP lexicon. Although much debated, this term refers to the output by agricultural businesses of non-market public goods and services that go beyond farming (OECD 2001). While uncertainty remains over what multifunctionality is and what purpose it serves, it has undoubtedly brought diversification to the centre stage of EU agricultural policy. Diversification of agricultural businesses usually increases the number of farm activities or functions, and, rightly or wrongly, EU policy-makers see it as instrumental in delivering the new multifunctional agriculture.

Clearly, then, the advent of the RDR and multifunctionality together offer a radical alternative to productivist agriculture – indeed, the first officially sanctioned alternative since the CAP's establishment in the late 1950s (Clark 2006). But what are the realities of agricultural diversification as a process? Can it deliver the material benefits claimed for it by the policy rhetoric, including increased business turnover, better utilization of on-farm assets and resources, and less reliance on public subsidies? And to what extent does it make new demands of agricultural businesses that are unsupported by existing public policy?

To answer these questions, we draw on empirical materials from an English case study. UK governments have been among the strongest advocates of CAP reform, seeking to reduce state intervention and to encourage diversification as part of a drive to reposition agriculture within rural economies. In England this has been catalysed by financial and public health crises (including BSE, classical swine fever and foot-and-mouth epidemics), resulting in a torrent of policy commissions and task force reports that identify the critical importance of agricultural businesses diversifying into new niche markets and supply chain initiatives

(Ministry of Agriculture, Fisheries and Food 2000, 2001; Cabinet Office 2002a, b; Countryside Agency 2004; Department of Environment, Food and Rural Affairs 2004). The English context, therefore, provides an interesting case where domestic upheaval in the sector has accompanied strategic policy change at EU level.

The chapter briefly outlines the main strands in research on agricultural diversification. We identify the prevalence of networks used in this work as (variously) a descriptive and analytical category. We then examine the extent to which the networks metaphor is reflective of concrete realities by scrutinizing farm business data collected in the English East Midlands region in 2001–2 as part of an international survey of agricultural diversification, funded by the European Commission. These data are used in two ways. The first is to assess the benefits of agricultural diversification in terms of business and regional economy effects. Second, we consider how far existing policy is supportive of agricultural diversification as a process, and consider alternative arrangements for regional governance. We finish by reviewing the possibilities for diversified agriculture in English rural regions in the future.

Agricultural diversification as a networked process

Geographers and agricultural economists have undertaken many valuable studies of agricultural diversification. Initially these identified a raft of explanatory factors, including on-farm resources, the skills and expertise of business managers, household succession, age profiles of farm managers, and so on as causal or contingent to diversification, without really referring to underlying process (e.g. Robson *et al.* 1987; Bromley and Hodge 1990; Fuller 1990; Brun and Fuller 1991). The impact of prevailing agricultural ideologies on these studies accounts for this emphasis on outcomes, thus explaining why 'theories of agricultural diversification … have yet to be fully developed' (Dorsey 1999: 179). So, from the 1970s to the mid-1980s in Western Europe, diversification was portrayed, rather unreflectively, as a static phenomenon, an adjunct to agricultural business turnover undertaken by a minority of managers without the resources of land, labour or capital to excel at 'farming' (Morgan and Munton 1971; Tarrant and Rex 1974; Haines and Davies 1987; Slee 1987).

The introduction of political-economy analyses into rural geography in the mid-1980s (with which Richard Munton was associated) challenged this view. Analysis from this perspective demonstrated agriculture's chronic structural supply–demand imbalance within Europe and globally. It also provided a theoretical context within which diversification could be addressed, focusing on the sector's peripherality to global capital accumulation (e.g. Marsden *et al.* 1987; Ilbery 1988), and the contribution made by off-farm diversification to business incomes (e.g. Whatmore 1991). Other researchers examined national and international approaches to de- and re-regulation and the resulting impact on agricultural diversification (Goodman and Watts 1997). Importantly, these studies clarified agriculture's interrelationships with other economic sectors, thereby promoting a better understanding of agro-food production and consumption. Implicit in many of these studies was the notion of networks of relations (for example, consideration of 'connections'/'interconnectedness'/'circuits' among actors and attributes), although this was rarely foregrounded.

From the mid-1990s onwards, however, rural geographers have engaged directly with diversification as a process. To do so they have used a variety of 'middle-range' concepts. Thus, examination has been made of the role of organizations in delivering agricultural diversification policies (Clark *et al.* 1997), while other researchers have invoked network theories, including actor networks (Whatmore and Thorne 1997; Murdoch *et al.* 2000), commodity chains (Bonanno 1994; Whatmore 1994) and agro-food networks (van der Ploeg *et al.* 2000; Renting *et al.* 2003), to analyse the scope and operation of diversified businesses and their supply side relations. The networks metaphor has attracted this attention because it enables research to move beyond depiction of agricultural diversification as interaction between spaces and actors 'internal' and 'external' to businesses, to more complicit understandings that trace the path of these interactions among and between actors, agencies and human–non-human attributes (Arce and Marsden 1993; Marsden and Arce 1995; Marsden 1998; Murdoch 2000; Kneafsey *et al.* 2001). These studies have usefully identified a wide range of network features and characteristics that impact on diversification and modify its outcomes (Table 10.1).

More recently, Marsden *et al.* (2002) have shown how networks can be used to understand the initiation and sustenance of agricultural diversification as a process. This means agricultural businesses developing '[the] new skills, new relationships and new entrepreneurialism required to participate in market relationships' (ibid.: 814). Specifically, businesses need 'to create and maintain new associations with a whole range of external actors and institutions' by 'constructing and optimising new social networks [with] those involved at different points in the various supply chains, and those significant regional and local actors who are able to facilitate [diversification]' (ibid.: 814–16).

Importantly, these and other authors have also identified the benefits accruing to diversifying businesses of this new networked agriculture. Philosophically, it enables managers to break away from the long-standing productivist logic of the

Table 10.1 Network characteristics and agricultural diversification

- Network type (what flows through the network, e.g. products, services, capital, information, knowledge, employment) (Curry and Winter 2000; Murdoch *et al.* 2000; Morris 2006)
- Spatial characteristics (e.g. horizontal ('in locality'), vertical ('out of locality'); singular or multi-nodal) (Murdoch 2000)
- 'Length' of networks (Marsden 1998; Renting *et al.* 2003)
- Nature of network (e.g. formal, informal, degrees of formality or informality, etc.) (Marsden and Arce 1995; van der Ploeg *et al.* 2000)
- Network cohesiveness/strength, diffuse/ephemeral (Jones and Clark 1997; Kneafsey *et al.* 2001)

CAP which has shaped the largely reactive configuration of European agriculture, geared to monolithic price-setting structures rather than capitalizing on emerging markets for public goods and quality foodstuffs. More pragmatically, it promises improved economic performance for businesses as the establishment and mobilization of new networks opens up hitherto unexploited local and regional markets and national and international demands.

Although unstated, the premise in all this work is that networking builds competitive advantage and so enhances the growth and survival prospects of farms. And, at the same time, diversifying businesses allegedly benefit local and regional economies by creating new employment and increasing cash flow. Clearly, there are many interconnected issues for researchers to resolve here, and in this chapter we offer only a preliminary analysis of two important questions. The first is to establish whether evidence of networking can be discerned among diversifying agricultural businesses; the second is to assess what benefit(s), if any, are realized by these businesses and their surrounding territories. The literature suggests that networks are instrumental to the collation and evaluation of knowledge by agricultural businesses, and the conversion of this knowledge into alternative enterprises that alter business structure. If so, it offers the prospect that agriculture can play a positive role in English rural futures. We examine this contention here with reference to agricultural business data from the East Midlands.

The English East Midlands

Fieldwork was conducted in England's East Midlands during 2001–2 (Figure 10.1). This region was selected for a number of reasons. First, its strong agricultural heritage meant that many actors and organizations had direct involvement in agricultural diversification. Second, farm businesses had a track record of diversifying (Farming and Rural Conservation Agency 1999), making identification and interviewing of business managers more straightforward. Third, the regional agricultural sector has traditionally been closely associated with wider regional development (e.g. through business-led processing and marketing initiatives), allowing examination of process and impacts to be undertaken.

Business addresses were obtained from regional offices of the former Ministry of Agriculture, Fisheries and Food, and comprised 500 agricultural enterprises (information comprised address, manager identity, business type and utilized agricultural area (UAA)). UAAs ranged from four to 400 hectares, and, in declining order of importance, arable, mixed arable–livestock and livestock holdings were all represented. A stratified sampling frame was adopted, and 118 (25 per cent) of these businesses agreed to participate in the survey. Data were collected from this sample through face-to-face semi-structured interviews with the manager of the agricultural business (that is, the person responsible for administration of the business, including diversified activities, if any). Where a business partner or spouse held responsibility for diversified activities, this individual was interviewed as well as the manager.

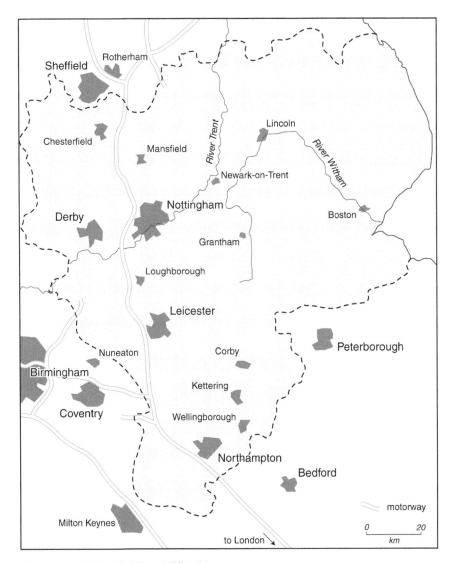

Figure 10.1 The English East Midlands.

Interviewees were asked to reflect on their business activities during 1997–2001 and identify any significant managerial changes made during this period, including diversification. Where diversification had been carried out, respondents were asked to consider whether networks were important to its incidence. The second part of the survey, relevant only to diversified businesses, examined the impact of diversification on agricultural businesses and local economies. We examine these aspects later. First, we profile the surveyed businesses.

Profile of surveyed agricultural businesses

The surveyed farms were mainly arable cropping enterprises made up of a mix of rented and owned land (in this respect, they were fully representative of farming across the region). The only business characteristic that discriminated diversified from non-diversified businesses was that the former were almost three times less likely to rent their land than the latter. This may reflect the difficulty tenant farmers have in diversifying their businesses, arising from legal stipulations in standard farm business leasehold agreements used in England and Wales. In demographic terms, too, there was no evidence of a difference between diversified and non-diversified businesses. The age distribution of the two groups was the same, they had similar family compositions and they had farmed for similar lengths of time.

Likewise, there were few significant differences in levels of educational attainment. However, a greater proportion of non-diversifiers had tertiary-level education, and they uniformly were more likely to have had an agrarian education. Among 35- to 45-year-olds, 90 per cent of non-diversifiers had a tertiary education, compared with 69 per cent of diversifiers. Some 77 per cent of non-diversifiers also had an agrarian education, compared with 55 per cent of diversifiers. This would seem to suggest that a tertiary-level agrarian education lessened the impetus to diversify. Perhaps most significant was the clear evidence that diversifiers uniformly had broader work experience *outside* farming than non-diversifiers.

Some 79 businesses (67 per cent of the sample) had diversified or were in the process of doing so, a process that, for most of them, started during the mid-1990s. There were 152 diversified activities distributed among these farms. The most commonly cited reasons for diversification were (in order): financial (19 per cent of respondents); employment retention or creation (18 per cent of respondents); to exploit emerging market opportunities (16 per cent); and to ensure the long-term survival of the business (16 per cent; diversifiers were also more strongly motivated by farm succession issues than non-diversifiers). As a management strategy, diversification addressed a wide range of business objectives, including survival, expansion, specialization and agricultural exit strategies.

These data also suggest that some diversifications were introduced to address short-term needs and problems, such as selling land to improve financial liquidity, while others (e.g. the refurbishment of buildings) were indicative of longer-term business planning and a change in the overall agricultural business enterprise. In 81 per cent of cases, responsibility for decision-taking was retained at the business level, with 86 per cent of all respondents either using existing capital or securing commercial loans to finance diversification, rather than using support from government organizations administering public funding specifically for this purpose.

Only ten out of the 118 business managers had 'never' considered diversification and had 'no plans' to do so, demonstrating the tremendous importance of this dynamic process to agricultural business strategies in the East Midlands.

Provision of non-agrarian services, including farm tourism, recreation, accommodation and catering, and handicrafts, was considered the most fruitful avenue for diversifying. For those who had considered diversification but subsequently abandoned the idea, the main explanation lay in their failing to secure planning permission, and inadequate assessment of market demand for the service or product they wanted to diversified into.

In order to realize on-farm diversification, businesses had, variously, adjusted their deployment of land and labour, and their use of knowledge and finance. During 1997–2001, 10 per cent of diversifiers increased their holding size, while 7 per cent decreased their holding size, possibly reflecting the instability in UK agricultural land markets at this time and the variable trends among farms to consolidate or liquidate their land assets. Some 16 per cent of diversifiers made significant changes in their level of on-farm employment, 11 per cent carried out a major overhaul of their marketing systems and 11 per cent altered their cropping mix in response to changing market demands. Diversified businesses were five times more likely to have replaced or refurbished farm buildings than non-diversifiers during the survey period, usually to provide or improve tourist accommodation. Significantly, off-farm employment provides an extremely minor component of income in comparison to that contributed by on-farm diversification.

Networks and the incidence of agricultural diversification in the East Midlands

The profile confirms the high prevalence of business diversification in the East Midlands and the wide spectrum of activities covered. Subsequent questions in the interview schedule sought to elucidate the processes underpinning these alternative activities. This showed that networks of relations and networking were causal factors in two areas, as follows.

Identifying and developing alternative business activities

Some 27 per cent of managers of diversified businesses had previous working experience beyond agriculture, compared to just 13 per cent of non-diversifiers, suggesting that diversified businesses had used existing 'out-of-sector' networks to identify their alternative business enterprises. This interpretation was corroborated by the finding that 44 per cent of non-diversifiers in the survey cited 'obstacles to networking' as the main reason for their failure to proceed. At the same time, 41 per cent of the diversified business sample had developed services and/or production wholly unrelated to agriculture, again suggesting the importance of 'out-of-sector' relations. This also emphasizes the openness of business managers to non-sectoral influences.

Mobilizing knowledge and resources for agricultural diversification

Some 76 per cent of managers of diversified businesses stated that networks of professional and/or informal contacts were 'essential' to some or all aspects of their diversification (e.g. to provide specialist advice, finance or regulatory guidance). Of these respondents, 44 per cent used networks to establish business opportunities or exploit market niches, confirming the importance of relations with other businesses and potential consumers in taking diversification forward. A total of 16 per cent of managers of diversified businesses also reported that they had developed new skills and/or developed new qualifications as a direct result of their networking activities, while 13 per cent had realized personal goals via networking. This implies that networks for agricultural diversification also offered social learning opportunities for some business managers. These networks appear to be used by diversifying businesses to mobilize resources that were not obtainable on-farm or in the immediate locality, and hence to sustain the on-farm diversification process, with 43 per cent of all diversifiers noting that diversification had required them to increase their network of social contacts.

Finally, many diversified businesses were disenchanted with, or actively disengaged from, the dominant productivist 'logic' of English farming. This was clear from the fact that while 43 per cent of respondents sought information from advisory bodies at the outset of the process, fewer than one in five contacted agricultural bodies specifically set up for this purpose. Some 82 per cent considered that the advice offered by these bodies was either not relevant or of no use to them, and 44 per cent believed that these organizations should provide more targeted advice.

These data give an impression of the role of networks in initiating diversification, but in order to explore this in more detail, interviewees were asked to reflect on the way(s) in which 'networking' and 'networked activities' had been useful or not to them. Although there was no such thing as a 'typical' network, certain structural similarities did emerge. Broadly, three types of networks were particularly important to business managers: those linking farm managers together; those bringing managers and regulatory bodies into contact; and those blending human and non-human attributes. We shall briefly discuss each of these in turn.

Networks bringing diverse 'outside' actors (including other farmers and consumers) and knowledge 'into' businesses were identified as very important by all diversifying business managers. Clearly, these networks allow them to compare their own situation to those of others who have already embarked on diversification and to appraise whether they have the resources necessary for success. As Table 10.1 suggests, these networks can have considerable geographical 'reach', and this is demonstrated in the following quotation from a farmer reflecting on how he started his spring-water bottling enterprise:

'If you listen, you can normally pick up a name, and often one name leads to another. I've spoken to a local farmer who's also got a spring. And he told me about this guy up in Scotland who's now set up a bottling plant.

And basically I went up to see him to pick his brains. I got advice on how he'd done it, and obviously that's better than trying to avoid the pitfalls yourself'.

In this case, network construction required the manager to draw down knowledge from different sources, confirming the contention of Marsden *et al.* (2002: 814) that 'new skills, new relationships and new entrepreneurialism [are] required by managers' for successful diversification. Only when networks have been created can mobilization of relations among actors and attributes begin. However, it is highly demanding to manage networks combining these often diverse regulatory, market and natural assets. In particular, managers need to 'optimiz[e] new social networks [with those] significant regional and local actors who are able to facilitate [diversification]' (ibid.: 816). This is an intensive, time-consuming process, often targeting individuals within specific organizations. It is well conveyed by a farm manager respondent's description of how he sought planning approval for reutilizing farm buildings. Tact, persuasion, negotiation and strategic design emerge as vitally important in building close relations with regulatory actors:

'[B]est is, to use your contacts, go to someone in the planning Department you know – or someone a friend knows. Take it [planning application] along informally; they'll have a look at it. They'll tell you if it's more or less OK: if it's not OK, then it's because of this, and if you change this, then it'll be OK. Then you go back, and you rewrite your application. You include their thoughts, ideas – it makes it harder for them to fail you. And when you meet again, hopefully they'll say, "yeah, that should work". Then there's a formal planning meeting but, well, you're up and running by then.'

All diversifiers in the survey showed some aptitude for combining resource assets and attributes in this way to ensure successful diversification. But arguably the most skilfully judged cases were where managers used natural phenomena and the resonances they provoke as an integral feature of their diversification. One interviewee commented on this aspect:

'[T]he ridge you see there, well it dominates the holding. It's a heft[2] for the livestock, it's rich in different grasses and shrubs and wildlife and it's where the spring originates. It's called Bluestone Heath, and I remember thinking that's a promising name for [the bottled mineral water he now sells]. And I knew Derbyshire Wildlife Trust were interested in the idea. And I thought if I could market it with their blessing, or even under their name, that'd bring *the landscape and the nature and the economy of the business together*, wouldn't it; it'd make for a real selling point' [emphasis added].

This melding of human-made and natural attributes also provides a means of reconnecting place through time. For example, one respondent's land was adjacent to a ruined monastery, creating a powerful sense of place that helped crystallize his alternative business:

> '[T]he sheep graze over the ruins of Thornton Abbey and that building's always held a strong association for us. And early on I did some reading and found that sheep's cheese was last made in this county at the time of the dissolution of the monasteries. And, chances are, one of the last places was right here. So I chose Thornton as the name of the hard cheese.'

Impact of agricultural diversification on businesses and the rural economy

These qualitative and quantitative results tend to support researchers' assertions about the importance of networks and networking by business managers to agricultural diversification. But what are the concrete results for diversifying businesses? In particular, is diversification capable of delivering fundamental change to agricultural businesses and territories, as is claimed by EU policy-makers?

Impacts on agricultural businesses

Results from the survey on this point were highly informative. Over the five-year survey period, 13 per cent of diversified businesses reported significant improvement in annual net business income as a direct result of their alternative enterprises, and 60 per cent stated that their income had remained constant or increased slightly, compared to just 4 per cent and 37 per cent respectively of non-diversified businesses. There were also positive employment benefits from agricultural diversification, with 54 per cent of diversified businesses employing one or more full-time workers and 49 per cent employing one or more part-time workers during 1997–2001. Another positive economic benefit was that these businesses were less dependent on CAP production subsidies. Some 58 per cent of diversifiers described their business as highly or very highly subsidy dependent, compared with 94 per cent of non-diversifiers.

Moreover, the survey reveals some interesting contrasts between the agricultural business incomes and future prospects of diversified and non-diversified businesses. For example, diversified farms were better off financially than non-diversified households, with 47 per cent earning in excess of £21,000 per year, compared with 36 per cent of non-diversified holdings. A total of 64 per cent of non-diversifiers earned less than £21,000 per year, compared with 53 per cent of diversifiers. And 60 per cent of diversifiers witnessed their household income staying the same or improving during 1997–2001, while almost a third (31 per cent) of non-diversifiers had seen their household incomes fall a lot over the same period. However, roughly a third of all diversifiers and a third of all non-diversifiers saw their level of borrowing increase slightly or a lot during 1997–2001. These economic effects have major consequences for business structure and for public policy. For example, 25 per cent of diversifications accounted for more than 50 per cent of total income of the relevant businesses, while 94 per cent of non-diversifiers described their businesses as highly or very highly dependent on the income contribution of agriculture and CAP subsidies, compared with just 58 per cent of diversifiers.

Unquestionably, diversified businesses had benefited financially. They experienced greater financial stability during the survey period, and had become less dependent on CAP subsidies to support total business income. Regrettably, establishing the precise level of business income was very difficult to assess, for three reasons. First, some alternative business activities were 'one-off', producing only a temporary benefit to the business (e.g. selling land or converting and selling a redundant building). Second, many diversifications had only been established for a short period; formation costs had been met, but income flows were only just beginning to rise. Third, some diversifications were effectively outside the business; indeed, there were cases in which they had effectively succeeded it. In general, though, diversified businesses in the East Midlands were very satisfied following the introduction of their alternative enterprise.

Impact on the East Midlands rural economy

Assessing the wider economic effects proved much more difficult. The survey identified a number of barriers that significantly reduced the impact of a network-based, diversified agriculture on the East Midlands economy. These related to the type of economic benefit derived from diversification, and the support structure facilitating diversification at the time the survey was undertaken. Where business diversification created employment, this tended to offset jobs lost through contraction of other parts of the business; diversification led to job preservation rather than job creation. The task of computing precisely how much new labour demand was created proved difficult because of the varied and complex structure of employment on East Midlands farms, a significant number of which (particularly on arable farms in the east of the region) employed several full- and/or part-time and/or casual workers, as well as seasonal gang labour. Overall, relatively few diversification initiatives actually created a demand for extra labour. Moreover, diversification often led to much longer working hours. A total of 54 per cent of all diversifiers noted that their alternative enterprises resulted in more work compared with previous arrangements. On average, non-diversified businesses put in 547 hours more per year into agricultural activities, suggesting that diversified businesses were spending several times this figure on their diversified activities annually.

Another problem was that the full economic benefits of diversification to the regional economy were not captured. Crucially, diversification was underdeveloped in terms of increasing value added, with the sample showing that existing agricultural skills, expertise and knowledge were used in almost 80 per cent of cases. Only 19 per cent of diversification activities depended on managers or their employees taking additional training or learning new skills. This finding was complicated, in that over a quarter (27 per cent) of the diversified group had previous business experience outside agriculture, compared to just 13 per cent of non-diversifiers. Hence, some of the skills and experience that diversified businesses attributed to being 'agricultural' or agriculturally based could instead have been learned through previous occupations.

Second, diversifiers across the region appear to have engaged in a very limited way with agricultural extension bodies. Thus, while 43 per cent of diversifiers sought help from public agencies, only 18 per cent joined funded schemes set up specifically to assist them. Part of the explanation appeared to be that these organizations did not offer diversifiers the resources they needed; hence, 54 per cent of all diversifiers said they needed a greater range of advice and information to diversify their businesses than was provided. Similarly, while diversified businesses in the East Midlands sought to develop 'extra-local' networks of association, ironically there appeared to be a marked reluctance to seek external sources of capital investment. Thus, 86 per cent of diversifiers relied on diverting finance from core farm business activities or on commercial loans to finance projects. Only 14 per cent of all diversifiers were in receipt of grant aid from external agencies or other funding bodies.

Finally, only 30 per cent of all diversifiers conducted market research before establishing their diversified enterprise, while just 28 per cent produced a business plan before proceeding to diversify. Once again this raises questions of whether agricultural organizations are providing the business management and administration skills required by diversifiers to tackle the often highly complex reskilling and reorganization tasks typically needed to bring diversification to a successful conclusion.

Analysis and consequences for public programmes and organizational delivery

The survey shows that diversified businesses derive clear economic benefits, though regional economic effects are far less apparent. Clearly, the success of these alternative enterprises is not being captured effectively enough to improve regional economic growth. One explanation seemed to be the relative disengagement between diversifiers and extension agencies in the region, which pointed to the need to review the support provided by regional governance to business managers. In particular, 'networked' diversification places onerous demands on managers in terms of their building, mobilizing and sustaining relations with a wide variety of people in extension and regulatory organizations. How far did support structures adequately address these demands? Could these be made more supportive to ensure that more agricultural businesses diversify?

To examine these questions, 68 semi-structured interviews were conducted with government agencies, county and local district councils, local politicians, small-business advice services, local rural development specialists, farm-level extension services, local community groups and private limited companies involved in diversification across the East Midlands. Specific organizations were identified by being cited by agricultural managers during the farm survey interviews. Wherever possible, semi-structured interviews were conducted with actors within the organization who had direct responsibility for diversification.

Based on their responses, and the theoretical approach adopted here of understanding diversification as a 'networked' process, we identified new roles that policy structures could play in the East Midlands. One of the most important findings to emerge from these interviews was that while sectoral and territorial actors play important roles in diversification, seemingly the involvement of *both* actor sets was critical to initiating and supporting forms of agricultural diversification with substantive employment generating effects. As a dynamic networked process, agricultural diversification in the region therefore appeared to rely upon different network forms at different times. So, some phases of the process cannot proceed without the involvement of policy practitioners closely associated with agriculture; other phases depend on the involvement of more informal non-sectoral contacts made socially; and some phases are reliant on both sectoral and social/territorial involvement to ensure success.

In identifying roles that policy structures could play, we combined these insights on sectoral and territorial actor involvement, and the critical importance of agricultural business 'networking' with them, with the need to address the shortcomings of the East Midlands policy structure identified from the business and institutional surveys. These shortcomings were:

1   There was no explicit regional strategy for agricultural diversification.
2   Consequently, there was no means of guiding the development of different diversifications (small, farm- or parish-based; larger, sectorally driven) in such a way as to enhance regional economic development outcomes.
3   No organization had direct responsibility for exhorting managers to pursue diversification, or to 'regionalize' policy by prioritizing certain diversification activities over others.
4   No organization had direct responsibility for facilitating the ad hoc policy structure around agricultural diversification.
5   No organization provided explicit support to the 'decision support structure' of businesses – that is, decision-making relating to diversification.

On this basis, we identify here two roles that organizations/policy structures could play regionally.

'Diversification champion'

Most respondents agreed that agricultural diversification would benefit from the introduction of a region-wide strategy for agricultural diversification. One commented:

'I'm not aware of any agency that's tasked with making diversification work in the East Midlands. And that, of course, comes back to there being no strategy to encourage diversification in these parts. I'm sure if you asked

DEFRA they could produce all sorts of bits of paper, but in practice there is very little, there is no real strategy to encourage change in this area.'

One approach would be for such a strategy to be developed by a prominent existing organization, such as the East Midlands Development Agency. This would offer a bridge between the differing aims and aspirations for diversification held by businesses, local communities and actors involved in regional economic development. A prime objective would be to prioritize forms of diversification that maintain or generate jobs in the rural economy. Certain diversification activities are better suited to this than others. As one interviewee observed:

'[A]t present, with the economic downturn, we've got farmers providing contracting services, but not setting up new businesses, and the money is really being kept in the agricultural sector, it's not seeping out into the local economy, not creating new jobs.'

Particular diversification activities should be prioritized that advance rural development goals and that do not lead to certain activities being over-represented, leading to market saturation. Diversification activities contributing to economic development could then be made attractive to farmers to encourage their uptake. The 'diversification champion' could also execute a variety of 'steering' functions to ensure the strategy's successful implementation, including setting targets for other organizations delivering diversification programmes and schemes; monitoring compliance among these institutions; setting standards for different diversification activities to drive quality upwards; and evaluating diversification outcomes and performance regionally.

Devising and implementing a diversification strategy, and ensuring that one organization is responsible for its execution, would enable this element of agricultural policy to be brought into the main stream of rural policy in the East Midlands. The 'diversification champion' could also provide a focus for establishing a working dynamic between local bodies and national government departments, including those with specific responsibilities in this area, such as DEFRA.

'Diversification supporter'

There is also a pressing need for one or a number of organizations to have an explicit front-line role in facilitating diversification locally by guiding agricultural business managers through the existing complex and often highly bureaucratic policy and administrative structure that supports it. Guidance should not be prescriptive but instead tailored to the needs of diversifiers, depending on their progress within a networked process with different stages, notably initiation, adjustment and sustenance (Clark 2005). An important role for these 'supporter' organizations would be to explain and facilitate policy-making and planning structures to diversifiers. The survey found that just 14 per cent of diversifiers were in receipt of grant aid, and only 14 per cent had joined relevant grant-aided schemes. Of those joining such schemes, 64 per cent did not find the advice provided by grant aid organizations useful.

While a range of communication technologies exist to disseminate this information (including the web, IT, mobile phone), the goal for 'supporters' would be to create a strong on-farm presence (the survey reported a lack of 'after-care' support – with only one in eight diversifiers received a personal visit from agencies). By gaining the trust of agricultural managers, and by working closely with the 'diversification champion', 'supporters' would be strongly placed to identify the diversification activity or activities most appropriate to farm businesses in particular localities. Another important function would be to identify potential business partners for agricultural businesses from among downstream suppliers within the food chain (the survey showed that only 1 per cent of diversifiers have external business partners). Supporter organizations could also elicit feedback from diversifiers on the transparency, administrative simplicity and usefulness of existing policy structures and programmes aimed at diversifying farms, thereby ensuring that their needs are reflected in the regional strategy.

Finally, organizational support would be even more effective if resources were available to promote diversification as a priority issue to agricultural business managers. 'Supporters' would be well placed to put diversifiers in touch with other farm businesses that had diversified into similar activities already, thus facilitating social learning and the transfer of knowledge between businesses. They would thus act more effectively as providers and disseminators of advice, and would be well placed to encourage potential managers to collaborate and possibly create partnerships, which many interviewees cited as a factor underlying diversification success. Consequently, these bodies would also be in a strong position to provide effective advice to business managers during different stages of diversification.

Building upon what has been gleaned from the empirical data on diversification as a process, these two roles consolidate existing organizational responsibilities regionally. But there are caveats regarding this approach. For example, while lip service is paid by the UK government to encourage agricultural diversification, no national guidance has been offered by DEFRA to its regional partners on this topic. The dilemma facing the region is thus one of cost-effectiveness: how to reach the greatest possible number of diversifiers while keeping transaction costs to a minimum?

Moreover, there appear to be conflicting allegiances among organizations in the East Midlands. A significant minority still promote the existing 'agricultural knowledge system', geared to maximizing food production. A critical question is how far this 'agricultural knowledge system' can encourage diversification into public goods production and service activities. As an interviewee from DEFRA commented:

'You've got a policy situation here that is clearly out of balance for the times – the incentive to stay in producing subsidized crops remains pretty high, and the bureaucracy and the application process associated with diversification means that only the really keen farmers will ever actually go for it.'

This issue of organizational allegiances affecting the region-wide development of diversification becomes particularly acute if one accepts the characterization of diversifying farmers as those who reject the ideology, technical trajectory and social structure of conventional farming (Kloppenburg 1991). A much clearer pledge by DEFRA to support diversification is needed, preferably made concrete by the implementation of a regionalized strategy for diversification along the lines discussed above.

Conclusions

Even for a region with a long history of agricultural diversification, policy initiatives in the East Midlands have developed in an ad hoc way. There is now a need for greater coordination and integration among and across sectoral and territorial organizations if the full economic benefits of diversified agriculture are to be reaped. Ideally, this should be driven by an explicit national strategy for diversified agriculture from DEFRA. The department needs to consult with its regional partners to determine their needs and priorities in this neglected area, and, where necessary, be prepared to release more resources to them via the English Rural Development Programme. If this approach were rolled out nationally, it would provide greater autonomy across English regions to fashion diversification strategies that were sensitive to local social, economic and environmental needs. Crucial to the development of this approach is recognition that diversification is a knowledge-intensive process, not a phenomenon, reliant on multidirectional flows of information between extension and regulatory organizations and agricultural businesses.

To some extent the neglect of diversification nationally reflects contrary impulses in the British state's administration of farming over the past decade. General recognition in Whitehall and Westminster of agriculture's diminished economic importance has fuelled the dominant impulse to reform the costly CAP and to make a national commitment to a more broad-based rural policy. However, this has been tempered over the past five years by a realization that the sector remains highly politically charged because of the real and symbolic consequences of food production. These include the very considerable public and financial impacts of recent disease epidemics and food 'scares', the implications of managing animal and zoonotic disease outbreaks (such as bird flu) in the future, and the possibility of agriculture reasserting its strategic importance as global food consumption begins to outpace production. And, at the same time, attempts by the state to divest itself of its long-standing responsibilities are complicated by the New Labour project of regionalization. In short, 'rolling back' state involvement in the sector is fraught with difficulties.

Faced with these contrary impulses, English policy has encouraged innovation and entrepreneurial business behaviour (diversification, in other words), while reducing domestic spending and subsidization (by cutting Treasury subventions in the sector, chiefly 'match funding' initiatives under the CAP). Paradoxically, the main casualty has been programmes such as those under the RDR, geared to encourage diversification.

In effect, over the past decade the state's post-war role as a provider of agricultural subsidy has moved uncertainly towards that of becoming a facilitator of sectoral management. An important element in this has been to promote multifunctionality as an organizing motif for agriculture, though, as the case study demonstrates, this has not been backed up by guidance from DEFRA to its regional partners regarding multifunctional aims and objectives, or with sufficient funds to make the multifunctional 'vision' a reality. The meagre finances made available to implement the recommendations of the Curry Commission on Farming and Food (Cabinet Office 2002b) typify the situation. Clearly, this will have implications for any sectoral transition from productivist to post-productivist positions in English agriculture. As Garzon (2005: 1), herself an official within the European Commission's Directorate General for Agriculture and Rural Development, notes, without resources it is only natural to ask, 'Is there [any] substance behind the [multifunctional] discourse's smoke?'

Crucially, then, the future contribution made by diversified agriculture to English regions depends on prevailing government policy. Assuming that this continues to be nationally applied and undifferentiated, encouraging rapid rural growth, small agricultural enterprises will decline further, and large institutional agribusinesses will focus on increasing productivity to match global demand. Diversification will remain an underexploited component of rural economic life, focused on speciality crops and niche products and carried out by a minority of entrepreneurial farmers. Alternatively, assuming that a more decentralized approach is introduced (including the increased resources, active support to businesses and enhanced strategic direction foreseen here), a more economically diverse picture could emerge. This would see locally developed solutions to the complex problem of linking alternative enterprise success with strong regional economic growth. Large and small agricultural businesses would benefit from bespoke guidance on diversification and environmentally sensitive farming practices within regional frameworks for public goods and services, including conservation, amenity and access. In effect, the current exhortations to diversify would be anchored to the reality of delivering a more sustainable future for England's rural regions.

Notes

1 This chapter focuses exclusively upon on-farm diversification.
2 An area of land with clear physical boundaries where livestock can be left to forage.

References

Arce, A. and Marsden, T.K. (1993) 'The social construction of international food: a new research agenda', *Economic Geography*, vol. 69, 293–311.

Bonnano, A. (1994) 'The locus of policy action in a global setting', in A. Bonnano, L. Busch, W. Friedland, L. Gouveia and E. Mingione (eds) *From Columbus to Con-Agra: The Globalization of Agriculture and Food*, Lawrence, KA: University of Kansas Press, pp. 251–64.

Bromley, D. and Hodge, I. (1990) 'Private property rights and presumptive policy entitlements: reconsidering the premises of rural policy', *European Review of Agricultural Economics*, 1990, vol. 17, 197–214.

Brun, A.H. and Fuller, A.M. (1991) *Socio-economic Aspects of Pluriactivity in Western Europe*, Enstone: Arkleton Trust.

Cabinet Office (2002a) *Scientific Review of Infectious Diseases in Livestock*, London: Cabinet Office.

Cabinet Office (2002b) *Farming and Food: A Sustainable Future. Report of the Policy Commission on the Future of Farming and Food*, London: Cabinet Office.

Clark, G., Bowler, I., Shaw, A., Crockett, A. and Ilbery, B. (1997) 'Institutions, alternative farming systems and local re-regulation', *Environment and Planning A*, vol. 29 731–45.

Clark, J.R.A. (2005) 'Examining the "new associationalism" in agriculture: agro-food diversification and multifunctional production logics', *Journal of Economic Geography*, vol. 5, 475–98.

Clark, J.R.A. (2006) 'The institutional limits to multifunctional agriculture: subnational governance and regional systems of innovation', *Environment and Planning C: Government and Policy*, vol. 24, 331–49.

Clark, J.R.A. and Jones, A.R. (2003) 'Introduction', in A.R. Jones (ed.) *Innovation, Diversification and European Agricultural Situations. Final Report to the European Commission*, Brussels: Commission of the European Communities, pp. 1–16.

Countryside Agency (2004) *Economies of Rural England: Reflecting the Realities*, Cheltenham: Countryside Agency.

Curry, N. and Winter, D.M. (2000) 'The transition to environmental agriculture in Europe: learning processes and knowledge networks', *European Planning Studies*, vol. 8, 107–21.

Department for Environment, Food and Rural Affairs (2004) *Rural Strategy 2004*, London: Department for Environment, Food and Rural Affairs.

Dorsey, B. (1999) 'Agricultural intensification, diversification and commercial production among small coffee growers in central Kenya', *Economic Geography*, vol. 75, 178–95.

Farming and Rural Conservation Agency (1999) *East Midlands Agricultural Issues: How They Affect the Rural Economy*, Lincoln: Farming and Rural Conservation Agency.

Fuller, A.M. (1990) 'From part-time farming to pluriactivity: a decade of rural change', *Journal of Rural Studies*, vol. 6, 361–73.

Garzon, I. (2005) 'Multifunctionality of agriculture in the European Union: is there substance behind the discourse's smoke?', Occasional Paper for the Institute of Governmental Studies (Agriculture and Resource Economics), Berkeley, CA: University of California at Berkeley.

Goodman, D.E. and Watts, M. (eds) (1997) *Agrarian Questions and Global Restructuring*, London: Routledge.

Haines, M. and Davies, R. (1987) *Diversifying the Farm Business*, Oxford: BSP Professional Books.

Ilbery, B.W. (1988) 'Farm diversification and the restructuring of agriculture', *Outlook on Agriculture*, vol. 17, 35–9.

Jones, A. and Clark, J.R.A. (1997) 'New directions in rural policy: a German perspective', *Built Environment*, vol. 23, 229–35.

Kloppenburg, J. (1991) 'Social theory and the de/reconstruction of agricultural science: local knowledge for an alternative agriculture', *Rural Sociology*, vol. 56, 519–48.

Kneafsey, M., Ilbery, B. and Jenkins, T. (2001) 'Exploring the dimensions of culture economies in rural west Wales', *Sociologia Ruralis*, vol. 41, 296–310.

Marsden, T.K. (1998) 'New rural territories: regulating the differentiated rural spaces', *Journal of Rural Studies*, vol. 14, 107–17.

Marsden, T.K. and Arce, A. (1995) 'Constructing quality: emerging food networks in the rural transition', *Environment and Planning A*, vol. 27, 1261–70.

Marsden, T.K., Banks, J. and Bristow, G. (2002) 'The social management of rural nature: understanding agrarian-based rural development', *Environment and Planning A*, vol. 34, 809–25.

Marsden, T.K., Munton, R.J.C. and Whatmore, S. (1987) 'Uneven development and the restructuring process in British agriculture', *Journal of Rural Studies*, vol. 3, 297–308.

Ministry of Agriculture, Fisheries and Food (2000) *Strategy for Agriculture: Current and Prospective Economic Situation*, London: Ministry of Agriculture, Fisheries and Food.

Ministry of Agriculture, Fisheries and Food (2001) *Rural Recovery Task Force Report*, London: Ministry of Agriculture, Fisheries and Food.

Morgan, W.B. and Munton, R.J.C. (1971) *Agricultural Geography*, London: Methuen.

Morris, C. (2006) 'Negotiating the boundary between state-led and farmer approaches to knowing nature: an analysis of UK agri-environment schemes', *Geoforum*, vol. 37, 113–27.

Murdoch, J. (2000) 'Networks: a new paradigm of rural development?', *Journal of Rural Studies*, vol. 16, 407–19.

Murdoch, J., Marsden, T.K. and Banks, J. (2000) 'Quality, nature and embeddedness: some theoretical considerations in the context of the food sector', *Economic Geography*, vol. 76, 107–25.

Organisation for Economic Co-operation and Development (2001) *Multifunctionality: Towards an Analytical Framework*, Paris: Organisation for Economic Co-operation and Development.

Renting, H., Marsden, T.K. and Banks, J. (2003) 'Understanding alternative food networks: exploring the role of short food supply chains in rural development', *Environment and Planning A*, vol. 35, 393–411.

Robson, N., Gasson, R. and Hill, B. (1987) 'Part-time farming: implications for farm family income', *Journal of Agricultural Economics*, vol. 38, 167–91.

Slee, B. (1987) *Alternative Farm Enterprises*, Ipswich: Farming Press.

Tarrant, J.R. and Rex, J. (1974) *Agricultural Geography*, Newton Abbot: David & Charles.

van der Ploeg, J.D., Renting, H., Brunori, G., Knickel, K., Mannion, J., Marsden, T.K., Roest, K., Sevilla-Guzman, E. and Ventura, F. (2000) 'Rural development: from practices and policies towards theory', *Sociologia Ruralis*, vol. 40, 391–408.

Whatmore, S. (1991) *Farming Women: Gender, Work and Family Enterprise*, London: Macmillan.

Whatmore, S. (1994) 'Global agro-food complexes and the refashioning of rural Europe', in A. Amin and N. Thrift (eds) *Globalization, Institutions and Regional Development in Europe*, Oxford: Oxford University Press, pp. 46–67.

Whatmore, S. and Thorne, L. (1997) 'Nourishing networks: alternative geographies of food', in D.E. Goodman and M. Watts (eds) *Globalising Food: Agrarian Questions and Global Restructuring*, London: Routledge, pp. 1–14.

11 Richard Munton

Geographer and rural geographer

Hugh Clout

This final chapter is written to celebrate the career of Richard Munton whose commitment to research and scholarship has made him a truly major figure in the world of rural geography and environmental conservation (Figure 11.1). Born in Bournemouth on 2 July 1943, Richard (John Cyril) received his secondary education at Poole Grammar School, where a succession of good

Figure 11.1 Richard Munton.

teachers kindled his interest in geography (Munton 2005). He was particularly attracted by the breadth of the subject, which encompassed such a wide range of concerns of relevance and political importance in the real world. Geography appeared to offer countless opportunities not found in other, more narrowly bounded disciplines, and hence Richard decided to take a degree in geography at university.

The Birmingham years

In the autumn of 1961 he began a three-year course at the University of Birmingham. The geography department comprised about a dozen academics and was headed by Professor David Linton, the distinguished geomorphologist (Gold *et al.* 1983). Three years earlier, at the age of 52, Linton had moved to Birmingham after 13 years as professor of geography at Sheffield. He was an excellent lecturer and was strongly committed to geographical education, working tirelessly on behalf of the Geographical Association. He was also a 'king-maker' whose advice was sought when academic promotions were to be made in British geography. The three other senior figures in the Birmingham department were Harry Thorpe, the historical geographer of the West Midlands; Gordon Warwick, the limestone geomorphologist; and Basil Johnson, the economic geographer, who would leave for a professorial post in Australia early in Richard's second year (Slater 1982, 1988).

These older men were joined by a group of younger lecturers, four of whom were particularly influential on Richard. Edward ('Ted') Stringer brought his experience of working at the Meteorological Office to teach meteorology and climatology, and was, of necessity, a quantifier before the so-called quantitative revolution impacted upon British geography. The three others were appointed by Linton and were still in their early thirties in 1961. William ('Bill') Morgan had spent six years lecturing at Ibadan in Nigeria, and in Birmingham taught agricultural geography and geographical ideas as well as Africa. Rowland Moss, appointed in Richard's second year at university, had spent five years as a soil scientist working for the government of Nigeria before becoming a lecturer at Ibadan. His expertise lay in ecology, biogeography and photo-interpretation, all of which he taught to Birmingham undergraduates, often in the context of Africa. Finally, there was Robin Donkin, who had spent a year with Professor Carl Sauer at the University of California, Berkeley, and who taught human and historical geography at Birmingham. Donkin's influence came in a very particular way. Richard admits that his mark on the first-year human geography examination paper (which was set for practice and returned, duly marked, to students) was, frankly, disappointing and he turned to the young examiner for advice. Whatever the problem may have been, Donkin set Richard on the right path that would lead to the career whose success we celebrate in this volume.

Beyond geography, Richard attended lectures by scholars from other departments, including the historians David Eversley and Douglas Johnson. The two men had impressive but very different teaching styles and contributed a whiff of the politics behind rational analyses of the world. Eversley was a fierce contro-versialist on local and regional planning matters, leading the 'Midlands New Towns Society' from the front as director of the West Midlands Social and Policy Research Unit. One of its main planks was defence of the Birmingham green belt. In their various ways, all these men shaped Richard's academic development, which resulted in the completion of an honours degree (Bachelor of Arts) in the summer of 1964.

However, by this time there was one special person whose love, companion-ship and family background would shape every aspect of Richard's subsequent life. She was a fellow geography student at Birmingham named Judith ('Judy') Morris, who would soon become Richard's wife. Judy was the daughter of a farmer in Shropshire, and gradually Richard became exposed to the practical world of farming life, having expressed an interested in the countryside and food production from a young age. In his own words:

'I don't come from the countryside and that was irritating to my father-in-law who didn't think a townie would have anything to say about agriculture. And after about twenty years he once agreed that I might know something about agriculture but I still didn't know anything about farming [in the prac-tical sense].'

(Munton 2005)

With a good degree behind him, Richard was invited to join the small group of postgraduates in geography at Birmingham in order to undertake doctoral research on British agriculture. Moss and Morgan would supervise this, and brought complementary expertise in biogeography and the methods of physical geography, and in economic geography and geographical ideas more widely. Richard's research involved numerous farm visits and countless interviews in fields and farmhouse kitchens, but it also involved relevant physical conditions, in particular soil conditions. Not surprisingly, given the influence of Rowland Moss, it was 'ecological' and 'environmental' in character, employing 'a lot of multivariate statistical analysis' before quantification became commonplace in British geographical practice (Munton 2005). Richard's earliest papers would demonstrate this 'logical-positivist quantitative' approach quite explicitly.

A career in London

With two years of his doctoral work completed, Richard turned to the question of future employment and responded to an advertisement for two assistant lec-tureships in economic geography placed by University College London (UCL). This was drawn to his attention by Peter Wood, at that time a graduate student in Birmingham, who would also join the staff at UCL and eventually become pro-fessor and head of department. One of these assistant lecturers would focus specifically on agricultural geography, which had previously been covered by J. Terry Coppock, who had left London to occupy the Ogilvie Chair of Geography in Edinburgh (Clout 2002). Richard was duly interviewed and was offered a post that would commence on 1 October 1966, the second appointee being John Salt from the University of Liverpool. The department of geography at UCL had been remodelled and enlarged during the 1950s and 1960s by Professor H.C. Darby (Clout 2003, forthcoming). By 1966 it contained two dozen academic staff and was the largest and widest-ranging department of geography in the United Kingdom, and was supported by the best-equipped cartography unit at that time. After a sabbatical break, Darby had left UCL in September 1966, passing the headship of the department to William ('Bill') Mead, who remained at the helm until 1981. This was a time of further expansion and also of rejuvenation as a number of staff who had been appointed by Darby moved to chairs and other senior appointments in other universities. Partly under Darby but especially under Mead, a cohort of young staff was appointed between 1965 and 1970, including, in date order, Peter Wood, Norman Perry, John Salt, Frank Carter, Hugh Clout, Carolyn Harrison, Bill Campbell, John Adams and Alan Gilbert as well, of course, as Richard. With just two exceptions, all would remain at UCL for their entire career, and of those who stayed, each rose to professorial status (Clout 2003: 30, 34).

Back in 1966, undergraduate teaching worked to a common syllabus that applied in all the colleges of the federal University of London, but change was not far away, and in 1968 each geography department in London introduced its own syllabus, 'course unit' and examining system. Not surprisingly, Richard's undergraduate lecturing focused on economic geography, especially agriculture, which would be reflected in his textbook *Agricultural Geography* (written with Bill Morgan, 1971). In addition to classes for undergraduates, Richard delivered postgraduate teaching in land-use planning and land economy to M.Sc. students on the Conservation course (then managed jointly with Botany, and to which Coppock had been strongly committed) and to M.Phil. students in Town Planning in the Bartlett School of Architecture and Planning at UCL. Over Richard's 40 years at UCL, these modules evolved in line with his developing research profile and acquired different titles. Thus, in 2000 he was providing intensive teaching on two Master's courses (Conservation; and Environment, Science and Society), a final-year undergraduate course on agricultural geography, half of a second-year course on environmental management and one-third of an introductory course on human ecology.

As well as teaching undergraduates and Master's students, Richard has supervised doctoral students at every stage of his career, beginning as early as 1969, only a year after finishing his own Ph.D. For a score of candidates he has been primary supervisor and for others he has operated in a secondary capacity along with Hugh Clout, Jacquie Burgess or Carolyn Harrison. The topics upon which he advised have broadened from agricultural geography of the British Isles to rural planning, conservation, recreation, coastal management and sustainable development (Table 11.1). All these students have been funded by studentships from the Economic and Social Research Council, with the exception of one seconded from the Canadian Ministry for the Environment, one funded by the States of Jersey and one Commonwealth Scholar. Unlike some academics,

Table 11.1 Doctoral students supervised by Richard Munton (date of award)

1976	C. Carr, 'A geographical analysis of the recruitment and outflow of young farm workers in eastern England, 1968–1972' (M.Phil.)
1978	A.M. Blair, 'Spatial effects of urban influences on agriculture in Essex, 1960–1974'
1979	P.C. Humphreys, 'Some structural and behavioural aspects of the adoption of the Small Farm (Incentive Bonus) Scheme in County Mayo, 1968–1974'
1981	J.G.M. Parkes, 'Coastal zone management in England and Wales: an analysis of the case for an integrated approach'
1984	S. Buchanan, 'Power and planning in rural areas: the preparation of the Suffolk County Structure Plan' (M.Phil.)
1986	E.G. Sharp, 'The acquisition of the Green Belt estates: a study in inter-authority relations'
1988	S.J. Whatmore, 'The "other half" of the family farm: an analysis of the position of "farm wives" in the familial gender division of labour on the farm'
1989	S. Pile, 'The deferential farmer: transformation and legitimation in advanced capitalist agriculture'
1990	S.A.W. Momtaz, 'Rural development and the problem of access: the case of the integrated rural development programme in Bangladesh'
1994	N. Ward, 'Farming on the treadmill: agricultural change and pesticide pollution'
1997	K.B. Collins, 'The politics of sustainable development: a case study of the island of Jersey'
1998	D.J. Rebane-Mortimer, 'Stitching the patchwork: an examination of the agri-environmental policy network in England'
1999	A.J. Jackson, 'Rural property rights and the survival of historic landed estates in the late twentieth century'
1999	K. Ahson, 'Innovation in agro-food biotechnology: a study in techno-science'
2001	D.A. Bloomfield, 'Towards the analysis of policy through the analysis of language: governance and brown field in London'
2003	T. Cooper, 'A critical examination of the social processes underpinning agricultural change: a case study of the viticultural sector, Languedoc'
2003	K. Studd, 'Fitness for purpose: English Nature's use of deliberative and inclusionary processes'
2004	S. Gardener, 'An evaluation of the effectiveness of stakeholder dialogue in environmental decision making'
2007	J. Mitchell, 'Public access, private space: the property fictions of the Countryside and Rights of Way Act 2002'
2007	I. Tomlinson, 'Transforming British organics: the role of central government, 1980–2006'

Richard rarely published with his doctoral students, 'because it's a tradition in the department that it's the student's thesis and not yours. But I sometimes wonder; I think of all the work that I have put into some of them, all stacked up there, those blue volumes' (Munton 2005) specially bound for the University of London.

The developing research profile

After he had moved to London, Richard's research career incorporated investigations into rural transport, the impact of recreational activities on local ecology, and then price trends in the agricultural land market, with special attention being paid to the role of financial institutions as owners of farm property. He was 'never told what to do' since 'it's basically a tradition of this department that you appoint staff who will develop their own research programme, and be the better for it' (Munton 2005). His methodology shifted 'into the political economy approach' involving the search for 'structures and politics and economics to explain change' in the countryside. The research on financial institutions led to Richard being appointed as an adviser from 1977 to 1979 to the Committee of Inquiry into the Acquisition and Occupancy of Agricultural Land established by the Ministry of Agriculture, Fisheries and Food under the chairmanship of Lord Northfield. Richard held special responsibility to report on the functioning of the rural land market and on investment decisions in agricultural land by financial institutions. In due course, Richard would collaborate with Robin Goodchild to write *Development and the Landowner: An Analysis of the British Experience* (1985) that in part drew on this experience. The book emphasized that there are many types of landowner and that they hold strikingly differing attitudes to the future development of their property. More importantly, the work on landownership began Richard's remarkable series of public policy commitments that continue to the present.

In 1977 he obtained a two-year award from the Department of the Environment for research into the management of agricultural land in the controlled environment of London's green belt. This gave rise to a cluster of academic papers and his monograph *London's Green Belt: Containment in Practice* (1983). This book traced the changing perceptions of green-belt land held by central and local government, especially the relations between restraint, amenity and land use, and analysed how local residents, councillors and recreation users continually refashion those perceptions to suit their own purposes. In 1978 Richard was co-recipient (together with Carolyn Harrison and Andrew Warren) of an award to analyse provision of land for informal recreation and the management priorities thereon. During the mid-1980s, Richard received six grants (from the Economic and Social Research Council, the Countryside Commission, the Council for the Protection of Rural England and the Leverhulme Trust) to investigate longitudinal relations between the ownership and management of rural land and the survival strategies of farming households. Empirical work traced the economic and social profiles of over 400 farm businesses and households in three lowland areas (London's green belt, west Dorset, east Bedfordshire) and two in the uplands (north Staffordshire, west Cumbria). In 1987 the Royal Geographical Society recognized this remarkable research activity by awarding Richard the Gill Memorial Award 'for contributions to the study of agriculture and land use in Great Britain'.

At this time he was developing an especially fruitful and friendly working relationship with Terry Marsden and Philip Lowe (as well as Jo Little, Jon Murdoch and Sarah Whatmore), which moved into an important new phase in 1987–8. The first step was the establishment in 1987 of the Rural Studies Research Centre at UCL under Richard's directorship. This comprised an interdisciplinary group of social scientists who were researching problems of rural development, largely in advanced economies. In Richard's words, with all this collaborative activity going on, 'it was a very exciting time' (Munton 2005). In May 1988 the Centre was successful in national competition when the Economic and Social Research Council selected it as one of two units (out of 30 bids) to receive core funding (in this case of £312,000) over four years to support its research. The resultant programme, headed by Richard Munton, Philip Lowe and Terry Marsden, was entitled 'The Social and Economic Restructuring of Rural Britain' and focused upon the land development process as a key mechanism in the social and economic restructuring of rural areas. It placed particular emphasis on the creation, contestation and reallocation of property rights between competing interests and how these reflected at local level the implementation of national policies for the countryside. A stream of multi-authored papers resulted, together with the important volume *Constructing the Countryside* (with T.K. Marsden, J. Murdoch, P. Lowe and A. Flynn, 1993), which linked developments in the global political economy to the behaviour of local people. This book and others by members of the team placed research in rural studies firmly in the mainstream of social science inquiry. Richard believes that these publications were 'probably the most influential' of all those with which he would be associated (Munton 2005).

The above research themes were complemented in another major research programme entitled 'Changing Farm Economies: Technological Change and Environmental Regulation' that was funded in 1989 under the Joint Agriculture and Environment Programme, which allocated £259,000 to Richard, and Philip Lowe. This study addressed the important issue of water pollution arising from modern agricultural practices (such as additions of nitrogen and pre-emergent cereal herbicides) and farm accidents (including leakage of slurry and silage effluent into watercourses). It sought to assess the likely responses of farmers to new technologies potentially less damaging to the environment and to new systems of environmental regulation, with the results being written up jointly with Neil Ward. This research led to Richard being appointed as an adviser on land-use questions to the House of Lords Special Select Committee inquiring into sustainable development, which critically reviewed the UK Government's *Strategy for Sustainable Development* (1994). Richard's colleague at UCL, Gerald Manners, was the other main adviser.

During the 1990s, Richard chaired the Project Directors Forum (1991–7) for the Community Forest Programme, served on the Forestry Commission's advisory panel on Urban and Community Forestry (1992–6), sat on the UK committee of the Man and Biosphere Committee (1992–9) and remains a member of English Nature's Socio-economic Advisory Group (1994–2006). In recent years Richard has responded to his wish to 'try something new' by focusing on the institutional and governance implications of policies directed towards sustainable development, including work on London's environmental governance (Munton 2005). As he reminded his students, 'I have always been interested in the political'. In addition, in 2002 Richard was appointed an adviser to the

Environment and Rural Affairs Department of the Scottish Executive on the future of socio-economic research. Over the years he 'learned more about politics and policy work from sitting on those committees than I ever learned really about the substance of the topics' (Munton 2005). Back in London, he was appointed director of the UCL Environment Institute (2003–6), which has the dual aim of encouraging interdisciplinary research into environmental issues in the college and promoting UCL's environmental research expertise to interests in government and business. This virtual centre has held meetings on cities and climate change, hosted major lectures by external speakers and organized seminars for graduate students and researchers.

Academic and citizen: a lifetime of service

Running parallel with this remarkable research career, which was recognized by his election to the Academy of Social Sciences (AcSS) in 2000, Richard served the Department of Geography at UCL, the college at large, the community of British geographers and the British research environment. After three years as assistant lecturer and 15 years as lecturer, he had been promoted in 1984 to a readership in recognition of his research achievements. After a further three years he was appointed to the established chair in Human Geography tenable at UCL, in succession to H.C. Darby and latterly to David Lowenthal. Richard used his experience of managing research to excellent effect, serving as coordinator of all research students in the department for three periods (1974–7, 1986–7 and 1998–9) and then rising to the exceptional challenge of being head of department from 1991 to 1997 (following the departure of Ron Cooke to be Vice-Chancellor of the University of York) and again from 2002 to 2005, in succession to Peter Wood. As head, Richard provided firm management for the continuing success of the department and its members, and was always an excellent source of informed and sympathetic advice. He adopted a very similar management style to that of Ron Cooke, ensuring that all significant documents were dealt with on his desk and chairing all important committees within the department (finance, examinations, resources, curriculum and strategy). He sought to combine top-down and bottom-up approaches, noting the essential paradox that 'the larger the organization, in some ways the more difficult it is to do bottom-up, but in some ways it is all the more important to do bottom-up', and to involve younger colleagues in management discussions (Munton 2005). In the wider context of UCL, Richard belonged, at various times, to two dozen committees, including the decisive Finance Committee (1975–8) when he was still a young man, the Graduate School Committee (1993–2001) and the important College Council (2004–6). As former Dean, I am particularly aware of his energetic contributions to the board and committees of the Faculty of Social and Historical Sciences, which developed from the Faculty of Arts in 1994.

Beyond UCL, Richard supported British geography in countless ways. Naturally, he served as external examiner on undergraduate degree programmes (in 11 universities) and examined doctoral candidates (in 15 institutions). In 1992 he took on the demanding task of working on the Geography Panel for the Research Assessment Exercise of the University Funding Councils, a role that he reassumed in 1995–6. On the next occasion (1999–2001), he was appointed chair of that critical panel and also chaired the Development Studies sub-panel (2000–1). As well as reading research submissions and convening the meetings of panel members, Richard was responsible for the delicate work of compiling the reports that revealed the grade achieved by each department entered for assessment. In undertaking this vital role, Richard arguably became the best-informed individual on the strengths and weaknesses of academic geography throughout the United Kingdom. On the strength of this unique expertise, he has been an external assessor (a 'king-maker', like Linton and Darby before him) for over 30 professorial chairs. Furthermore, he conducted formal reviews of 18 departments of geography between 1997 and 2006, with more to come.

Within the Royal Geographical Society with the Institute of British Geographers (IBG), Richard served between 1969 and 1972 as honorary secretary of the Agricultural Study Group, which had been initiated by Terry Coppock. He was a committee member of its successor, the Rural Geography Study Group, in 1984–5 and was elected chair of that group for the following three years, during which he provided enthusiastic leadership. Together with Andrew Goudie, he responded to an invitation from the British National Committee for Geography of the Royal Society to write a 'Review of British geography 1980–1984' in advance of the International Geographical Congress held in Paris during August 1984 (see the Appendix). Geographers from all parts of the world received copies of this important document. In 1994–5 Richard was vice-chair of the Environmental Research Group and subsequently became chair. After being Vice-President of the RGS/IBG during 1997–9, he was chair of the Annual Conference in January 1999. After having been a member of Council of the then IBG from 1984 to 1987, he remains an active member of the research and higher education committee and of the research advisory committee of the RGS/IBG.

Beyond the disciplinary boundaries of geography, in the early 1980s Richard co-edited (with Judith Rees) the Resource Management Series of research monographs for publisher George Allen & Unwin. Towards the end of that decade he was a consultant to the joint research council working group on agriculture and the environment established by the Advisory Board for the Research Councils (1987–8). He has served on four subcommittees of the Economic and Social Research Council, and between 1992 and 1998 chaired the steering group of a major research interdisciplinary programme called 'The Nation's Diet', funded by the ESRC. This, he believed, 'was a pretty imaginative step when it was set up back in the 1990s', before obesity became such a pertinent topic in the United Kingdom (Munton 2005). Richard has been active, too, on committees of the Natural Environment Research Council, chairing the Land-Use Research Committee (1991–4) and the Land-Use Research Steering Committee (1994–7). From 1998 to 2003 he sat on the programme management committee of the link programme on Eating, Food and Health of the Biology and Biotechnology Research Council. Richard was on the executive committee of the

Association of Learned Societies in the Social Sciences (1984–8) and then coordinated, on behalf of ALSISS, the seminar programme for the All-Party Group of the Houses of Parliament on Social Science and Policy (1989–95). In 2003 he was appointed chair of the advisory committee to the Sustainable Development Research Network run by the Policy Studies Institute and funded by the Department for Environment, Food and Rural Affairs. For Richard, 'sustainable development is about a process of change where hopefully environmental issues are given full credit … in relation to social and economic considerations', a process where 'environmental issues have to be taken seriously in terms of the future' (Munton 2005). In the following year he was elected to the Council of the Academy of Social Sciences (2004–). Over and above all these activities, Richard has found time to act as UCL representative on the court of the University of Birmingham (1995–2006) and is the University of London governor on the board of governors of St Albans School (1998–), which is not far from his home in Hertfordshire.

Into the future

Without doubt, Richard's commitment to geography, the social sciences, public policy, UCL and numerous other organizations and individuals will not cease upon formal retirement from UCL at the end of September 2006. He will surely continue to be called upon to share his rich expertise in so many areas and, we know, is keen to maintain his close contacts with graduate students and researchers working on rural geography and, more broadly, on environmental themes. Retirement should, however, grant him more time for himself and for members of his family, for whom geography has been such a central part of their lives. Not only was Judy a teacher of geography throughout her career but their children, David and Anne, both took degrees in geography. It is too early to report on the academic choices of the grandchildren. As well as being a statement of recent research in rural geography, this volume is an expression of thanks to Richard from his colleagues and friends in the overlapping worlds of geography and UCL. Each contributor is indebted to him for leadership, encouragement and support. We join together in wishing Richard and Judy many happy years ahead.

References

Clout, H. (2002) 'John Terence Coppock 1921–2000', *Proceedings of the British Academy*, vol. 115, 207–24.

Clout, H. (2003) *Geography at University College London: A Brief History*, London: UCL.

Clout, H. (forthcoming) 'Henry Clifford Darby 1909–1992', *Geographers: Biobibliographical Studies*.

Gold, J.R., Haigh, M.J. and Warwick, G.T. (1983) 'David Leslie Linton 1906–1971', *Geographers: Biobibliographical Studies*, vol. 7, 75–83.

Munton, R., Transcript of an interview by first-year students James Hampson, James

Hiatt and Adam Dobson as part of the staff research project for the Writing and Analysis in Geography module, March 2005.

Slater, T.R. (1982) 'Professor Harry Thorpe, 1913–1977: an appreciation and assessment', in T.R. Slater and P.J. Jarvis (eds) *Field and Forest: An Historical Geography of Warwickshire and Worcestershire*, Norwich: GeoBooks, pp. 9–30.

Slater, T.R. (1988) 'Redbrick academic geography', *Geographical Journal*, vol. 154, 169–80.

Appendix

Books, refereed articles, book chapters and other publications and reports by Richard Munton listed in chronological order

Books

1971   (with W.B. Morgan) *Agricultural Geography*, London: Methuen, 157 pp.
1983   *London's Green Belt: Containment in Practice*, London: George Allen & Unwin, 190 pp.
1985   (with R.N. Goodchild) *Development and the Landowner: An Analysis of the British Experience*, London: George Allen & Unwin, 220 pp.
1993   (with T.K. Marsden, J. Murdoch, P. Lowe and A. Pratt) *Constructing the Countryside*, London: UCL Press, 220 pp.

Refereed articles

1969   (with J.M. Norris) 'The analysis of farm organization: an approach to the classification of agricultural land in Britain', *Geografiska Annaler Series B*, vol. 52, 95–103.
1969   (with J. Salt) 'The Redcliffe–Maud Report: the countryside', *Area*, vol. 4, 18–20.
1970   (with B. Goldsmith and A. Warren) 'The impact of recreation on the ecology and amenity of semi-natural areas: methods of investigation used in the Isles of Scilly', *Biological Journal of the Linnean Society*, vol. 2, 287–306.
1971   (with H. Clout) 'The problem bus', *Town and Country Planning*, vol. 39, 112–16.
1973   'Recent trends in farmland prices in England and Wales', *Estates Gazette*, vol. 227, 2159–65.
1975   'The state of the agricultural land market 1971–73: a survey of auctioneers' property transactions', *Oxford Agrarian Studies*, NS, vol. 4, 111–30.
1976   'An analysis of price trends in the agricultural land market of England and Wales', *Tijdschrift voor Economie en Sociale Geografie*, vol. 67, 202–12.
1977   'The financial institutions: their ownership of agricultural land in Great Britain', *Area*, vol. 9, 29–37.
1979   (with M.J. Ferguson) 'Informal recreation sites in London's Green Belt', *Area*, vol. 11, 196–205.
1981   'Management expenditure on informal recreation sites in London's Green Belt', *The Planner*, vol. 67, 93–5.
1981   'Agricultural land use in the London Green Belt', *Town and Country Planning*, vol. 49, 17–19.
1981   'Agricultural land tenure in Great Britain: some current issues', *Revue d'Économie Régionale et Urbaine*, vol. 3, 291–303.

1982   'The university and rural resource development', *Sociologia Ruralis*, vol. 22, 57–62.

1983   'Green Belt policy: what role for agriculture?', *Planning Outlook*, vol. 25, 43–51.

1984   (with A. Goudie) 'Review of British geography 1980–1984', *Geographical Journal*, vol. 150, 27–47. [Following an invitation from the British National Committee for Geography of the Royal Society, and prepared for the International Geographical Union Congress, Paris, August 1984]

1985   'Investment in British agriculture by the financial institutions', *Sociologia Ruralis*, vol. 35, 155–73.

1986   'Green Belts: the end of an era?', *Geography*, vol. 71, 206–14.

1986   (with T.K. Marsden, J.K. Little and S.J. Whatmore) 'Towards a political economy of capitalist agriculture: a British perspective', *International Journal of Urban and Regional Research*, vol. 10, 498–521.

1986   (with T.K. Marsden, S.J. Whatmore and J.K. Little) 'The restructuring process and economic centrality in capitalist agriculture', *Journal of Rural Studies*, vol. 2, 271–80.

1986   (with S.J. Whatmore, T.K. Marsden and J.K. Little) 'Internal and external relations in the transformation of the family farm', *Sociologia Ruralis*, vol. 26, 396–98.

1987   (with S.J. Whatmore, T.K. Marsden and J.K. Little) 'Towards a typology of farm businesses in contemporary British agriculture', *Sociologia Ruralis*, vol. 27, 21–37.

1987   (with S.J. Whatmore, T.K. Marsden and J.K. Little) 'Interpreting a relational typology of farm businesses in southern England', *Sociologia Ruralis*, vol. 27, 103–22.

1987   (with T.K. Marsden and S.J. Whatmore) 'Uneven development and the restructuring process in British agriculture: a preliminary exploration', *Journal of Rural Studies*, vol. 3, 297–308.

1988   (with S.J. Whatmore and T.K. Marsden) 'Reconsidering urban-fringe agriculture: a longitudinal analysis of capital restructuring on farms in the metropolitan green belt', *Transactions of the Institute of British Geographers*, vol. 13, 324–36.

1989   (with T.K. Marsden and S.J. Whatmore) 'Part-time farming and its implications for the rural landscape: a preliminary analysis', *Environment and Planning A*, vol. 21, 523–36.

1989   (with T.K. Marsden and S.J. Whatmore) 'Strategies for coping in capitalist agriculture: an examination of the responses of farm families in British agriculture', *Geoforum*, vol. 20, 1–14.

1990   (with S.J. Whatmore and T.K. Marsden) 'The rural restructuring process: emerging divisions of agricultural property rights', *Regional Studies*, vol. 24, 235–45.

1990   (with N. Ward and T.K. Marsden) 'Farm landscape change: trends in upland and lowland England', *Land Use Policy*, vol. 7, 291–302.

1991   (with T.K. Marsden) 'Dualism or diversity in family farming? Patterns of occupancy change in British agriculture', *Geoforum*, vol. 22, 105–17.

1991   (with T.K. Marsden) 'Occupancy change and the farmed landscape: an analysis of farm-level trends, 1970–1985', *Environment and Planning A*, vol. 23, 499–510.

1991   (with T.K. Marsden) 'Occupancy change and the farmed landscape: implications for policy', *Environment and Planning A*, vol. 23, 663–76.

1992   (with N. Ward) 'Conceptualising agriculture–environment relations: combining political economy and socio-cultural approaches to pesticide pollution', *Sociologia Ruralis*, vol. 32, 21–38.

1992   (with T.K. Marsden and N. Ward) 'Incorporating social trajectories into uneven agrarian development: farm businesses in upland and lowland Britain', *Sociologia Ruralis*, vol. 32, 408–30.

1993   (with P. Lowe, J. Murdoch, T.K. Marsden and A. Flynn) 'Regulating the new rural spaces: the uneven development of land', *Journal of Rural Studies*, vol. 9, 205–22.

1995   'Regulating rural change: property rights, economy and environment: a case study from Cumbria, England', *Journal of Rural Studies*, vol. 11, 269–84.

1996   (with T.K. Marsden, N. Ward and S.J. Whatmore) 'Agricultural geography and the political economy approach: a review', *Economic Geography*, vol. 72, 361–75.

1997   'Engaging sustainable development: some observations on progress in the UK', *Progress in Human Geography*, vol. 21, 147–63.

2001   (with D. Bloomfield, K. Collins and C. Fry) 'Deliberation and inclusion: vehicles for increasing trust in UK public governance?', *Environment and Planning C*, vol. 19, 501–13.

2004   (with C.M. Harrison and K. Collins) 'Experimental discursive spaces: policy processes, public participation and the Greater London Authority', *Urban Studies*, vol. 41, 903–17.

Chapters in books

1969   'The economic geography of agriculture', in R.U. Cooke and J.H. Johnson (eds) *Progress in Geography*, Oxford: Pergamon Press, 143–52.

1971   (with B. Goldsmith) 'The ecological effects of recreation', in P. Lavery (ed.) *Recreational Geography*, Newton Abbot: David & Charles, 259–69.

1972   'Farm systems classification: a use of multi-variate analysis', in *Agricultural Typology and Land Utilisation (IGU Commission on Agricultural Typology)*, Verona, 89–107.

1973   'Systems analysis: a comment', in C. Renfrew (ed.) *The Explanation of Cultural Change*, London: Duckworth, London, 685–90.

1974   'Farming in the urban fringe', in J.H. Johnson (ed.) *Suburban Growth: Geographical Processes at the Edge of the City*, Chichester: Wiley, 201–23.

1974   'Agriculture and conservation in lowland Britain', in B. Goldsmith and A. Warren (eds) *Conservation in Practice*, Chichester: Wiley, 323–36.

1978   'London's Green Belt', in H.D. Clout (ed.) *Changing London*, Cambridge: University Tutorial Press, 99–109.

1979   'The financial institutions: their interests in farmland', Appendix 5 to the *Report of the Committee of Inquiry into the Occupancy and Acquisition of Agricultural Land*, Cmnd. 7599, London: Her Majesty's Stationery Office, 304–39.

1983   'Agriculture and conservation: what room for compromise?', in B. Goldsmith and A. Warren (eds) *Conservation in Perspective*, Chichester: Wiley, 353–73.

1984   'Land speculation and the under-use of urban-fringe farmland in the Metropolitan Green Belt', in G. Clark, J. Groenendijk and F. Thissen (eds) *The Changing Countryside*, Norwich: GeoBooks, 221–32.

1984   'The politics of rural land ownership: institutional investors and the Northfield enquiry', in A. Bradley and P. Lowe (eds) *Locality and Rurality: Economy and Society in Rural Regions*, Norwich: GeoBooks, 167–80.

1986 'The Metropolitan Green Belt', in H. Clout and P. Wood (eds) *London: Problems of Change*, London: Longman, 128–36.

1987 'The conflict between conservation and food production in Great Britain', in C. Cocklin, B. Smith and T. Johnston (eds) *Demands on Rural Lands*, Boulder, CO: Westview Press, 47–60.

1987 (with J. Eldon and T.K. Marsden) 'Farmers' responses to an uncertain policy future', in D. Baldock and D. Conder (eds) *Removing Land from Agriculture: The Implications for Farming and the Environment*, London: IEED/CPRE, 19–30.

1990 (with S.J. Whatmore and T.K. Marsden) 'The role of banking capital and credit relations in British food production', in T.K. Marsden and J. Little (eds) *Political, Social and Economic Perspectives on the International Food System*, Aldershot: Avebury, 36–56.

1990 (with T.K. Marsden and S.J. Whatmore) 'Technological change in a period of agricultural adjustment', in P. Lowe, T.K. Marsden and S. Whatmore (eds) *Technological Change and the Rural Environment*, London: Fulton, 104–26.

1990 (with P. Lowe, G. Cox, D. Goodman and M. Winter) 'Technological change, farm management and pollution regulation', in P. Lowe, T.K. Marsden and S. Whatmore (eds) *Technological Change and the Rural Environment*, London: Fulton, 53–80.

1992 (with P. Lowe and T.K. Marsden) 'Forces driving land use change: the social, economic and political context', in M. Whitby (ed.) *Land Use Change: The Causes and Consequences*, ITE Symposium No. 27, London: Her Majesty's Stationery Office, 15–27.

1992 (with P. Lowe and N. Ward) 'Social analysis of land use change: the role of the farmer', in M. Whitby (ed.) *Land Use Change: The Causes and the Consequences*, ITE Symposium No. 27, London: Her Majesty's Stationery Office, 42–51.

1992 'The uneven development of capitalist agriculture: repositioning within the food system', in K. Hoggart (ed.) *Agricultural Change, Environment and Economy: Essays in Honour of W.B. Morgan*, London: Mansell, 25–48.

1992 (with T.K. Marsden and N. Ward) 'Uneven agrarian development and the social relations of farm households', in I. Bowler, C. Bryant and D. Nellis (eds) *Contemporary Rural Systems in Transition*, Oxford: CAB International, 61–73.

1992 'Land, labour and capital', in I. Bowler (ed.) *The Geography of Agriculture in Developed Market Economies*, London: Longman, 56–84.

1997 'Sustainable development: a critical review of rural land-use policy in the UK' in B. Ilbery, Q. Chiotti and T. Rickard (eds) *Agricultural Restructuring and Sustainability*, Oxford: CAB International, 11–24.

2003 'Deliberative democracy and environmental decision making', in F. Berkhout, M. Leach and I. Scoones (eds) *Negotiating Change: Advances in Environmental Social Science*, Camberley: Edward Elgar, 56–82.

2006 (with D. Goode) 'The Greater London Authority and Sustainable Development', in M. Tewdwr-Jones and P. Allmendinger (eds) *Territory, Identity and Spatial Planning: Spatial Governance in a Fragmented Nation*, London: Routledge, 237–54.

Other publications

1969 'Agricultural land classification and rural planning', *Chartered Surveyor*, 96–7.

1972 (with H.D. Clout) 'The geographical implications of changing patterns of personal mobility for the spatial organization of central places in rural Norfolk (East Anglia, England)', in B. Sarfalvi (ed.) *Urbanization in Europe* (International Geographical Union Regional Conference), Budapest: Hungarian Academy of Sciences, 167–81.

1974   (with J. Rees) *Resource Management Register: A Survey of Current Geographical Work in Resource Management*, London: Social Science Research Council, 57 pp.

1975   'Agricultural land price survey in England, 1971–1973: some preliminary results', *Chartered Surveyor, Rural Quarterly*, vol. 2, 59–64.

1976   'Agricultural land prices in 1974: some observations', *Chartered Surveyor, Rural Quarterly*, vol. 4, 14–16.

1980   'The Northfield Report: a comment', *Countryside Planning Yearbook*, vol. 1, 86–9.

1982   (with G. Goransson, W.J. Kimball and U. Renborg) *The University and Rural Resource Development: Conference Proceedings*, Uppsala: Swedish University of Agricultural Sciences, 66 pp.

1982   'The Northfield Committee Report and small farms', in B.J. Marshall and R.B. Tranter (eds) *Small Farming and the Rural Community*, Reading: University of Reading Centre for Agricultural Strategy, 18–27.

1983   'Agriculture: a review of 1981–2', *Countryside Planning Yearbook*, vol. 4, 159–62.

1983   'Resource management', *Progress in Human Geography*, vol. 7, 126–32.

1984   'Green belt and land for housing', Evidence submitted to the Environment Select Committee of the House of Commons in its inquiry into the above subject, Vol. III, London: Her Majesty's Stationery Office, 121–32.

1984   'Resource management and conservation: the UK response to the World Conservation Strategy', *Progress in Human Geography*, vol. 8, 120–6.

1985   'Resource management: the problem of value', *Progress in Human Geography*, vol. 9, 264–70.

1987   'Research in rural geography in Britain: some reflections on future trends', *Netherlands Geographical Studies*, vol. 27, 30–40.

1987   (with J. Groch and C. Harrison) 'Funkcja wypoczynkowa podmiejskiej strefy Londynu (London's Green Belt)', *Prace Geograficzne*, vol. 68, 87–96.

1988   'Thatcher's countryside: agriculture and the rural economy', *International Yearbook of Planning and the Countryside*, vol. 2, 39–43.

1988   (with T.K. Marsden and J. Eldon) *Occupancy Change and the Farmed Landscape*, CCD 33, Cheltenham: Countryside Commission, 19 pp.

1989   'Londres: ceinture verte et rocade', *Études Foncières*, vol. 44, 24–30.

1989   'The context and objectives of research centre management: background paper', in R. Walker and P. Stringer (eds) *Managing Interdisciplinary Research* (Proceedings ESRC Seminar, University of York), Swindon: Economic and Social Research Council, 12–14.

1989   (with P. Lowe and T.K. Marsden) 'The social and economic restructuring of rural Britain: A Position Statement', Working Paper 2, Countryside Change Series, 24 pp.

1990   (with P. Lowe, T.K. Marsden and N. Ward) *Memorandum on the Future of Rural Society: Select Committee on the European Communities, House of Lords, HL 80–111*, London: Her Majesty's Stationery Office, 460–70.

1990   (with T.K. Marsden and S. Whatmore) 'Recent changes in British agriculture (Appendix E), *Report of the Archbishop's Commission on Rural Areas* (*Faith in the Countryside*), Worthing: Churchman, 361–8.

1990 (with J. Murdoch and J. Grove-Hills) 'The rural land development process: evolving a methodology', Working Paper 8, Countryside Change Series, 32 pp.

1990 (with A. Jones) 'Set-aside story', *Geographical Analysis, Geographical Magazine*, December, 1–3.

1991 'The social and economic restructuring of rural Britain', *Progress in Rural Policy and Planning*, vol. 1, 106–8.

1994 'Classics in human geography revisited: the case of Chisholm's "Rural Settlement and Land Use"', *Progress in Human Geography*, vol. 18, 59–60.

1996 (with S.J. Whatmore, T.K. Marsden and J. Little) 'The trouble with subsumption and other rural tales: a response to critics', *Scottish Geographical Magazine*, vol. 112, 54–6.

1997 'The land of Britain: its use and misuse', in A. Mackay and J. Murlis (eds) *Britain's Natural Environment: A State of the Nation Review*, London: ENSIS, 19–22.

1998 (with K. Collins) 'Government strategies for sustainable development', *Geography*, vol. 83, 346–57.

1999 (with J. Burgess, K. Collins, C.M. Harrison and J. Murlis) *An Analytical and Descriptive Model of Sustainable Development for the Environment Agency*, Environment Agency Report Series, 13, Environment Agency: Swindon, 77 pp.

Research reports and unpublished evidence

1986 *Land Occupancy and Amenity in the Urban Fringe*, Final Report to the Economic and Social Research Council.

1986 (with T.K. Marsden) Evidence submitted to the Countryside Commission's Review Panel on Countryside Policy.

1987 (with T.K. Marsden) *Occupancy Change and the Farmed Landscape*, Final Report to the Economic and Social Research Council.

1987 (with T.K. Marsden and J. Eldon) *Occupancy Change and the Farmed Landscape*, Report submitted to the Countryside Commission.

1987 *Rural Economy and Society*, Report prepared for the Joint Party Working Group of the Advisory Board for the Research Councils on Agriculture and the Environment.

1987 (with T.K. Marsden and S. Whatmore) *External Capitals and Their Role in Occupancy Change in the Farmed Landscape*, Report to the Economic and Social Research Council.

1988 (with D. Rhind, D. Briggs and J. Townshend) Evidence submitted to the Butterworth Committee on The Future of Agricultural Research, House of Lords Select Committee on Science and Technology, prepared on behalf of the Royal Geographical Society and Institute of British Geographers.

1989 (with T.K. Marsden and N. Ward) *Social and Economic Change in Upland Agriculture: Case Studies from Cumbria and North Staffordshire*, Report to the Leverhulme Trust.

1993 (with P. Lowe and T.K. Marsden) *The Social and Economic Restructuring of Rural Britain*, Final Report to the Economic and Social Research Council.

Index